A CO-REQUISITE WORKBOOK FOR
STATISTICS

PELLISSIPPI STATE
COMMUNITY COLLEGE
MATHEMATICS FACULTY

Kendall Hunt
publishing company

TABLE OF CONTENTS

FOREWORD

Co-requisite remediation—as opposed to the traditional, pre-requisite (sequential) model of remediation—seeks to help advance under-prepared college students by providing additional academic support and removing common attrition points while enrolled in a gateway, college-level course. The Tennessee Board of Regents, which governs all 13 public community colleges in the state, was one of the first systems of higher education to fully adopt to scale the co-requisite approach to Mathematics, English, and reading. Preliminary results have been encouraging, with the percentage of students completing a college-level math course within two years increasing from 12% in 2014 to 61% in 2015.

Remedial mathematics is decidedly "algebra-heavy," with topics like polynomial multiplication and factoring that are of little use to a non-STEM student. In particular at Pellissippi State Community College, over 80% of majors offered by the College require or accept Introductory Statistics (non-calculus based) as credit for a mathematics course towards graduation. This has led to the creation of a specialized curriculum for the co-requisite course. Specifically, the context of all the content in this workbook is geared directly to needs in a first-year introduction to statistics course. It is designed to provide *just in time* remediation, with instruction that is directly tied to statistical concepts. In this way, the authors felt that students would see the co-requisite course as a direct aid to their success in the statistics course, rather than simply another math course in addition to the college-level course.

The workbook structure is arranged according to the typical progression of an introductory statistics and probability course and could be used to support a variety of textbooks. Some topics, such as the order of operations and evaluating expressions, are revisited throughout the co-requisite course, progressing from simplistic formulas such as the sample mean, to more complex expressions, such as the margin of error and test statistics. The title heading of each unit and lesson should allow professors to pick and choose workbook sections which might be of help to individual students.

UNIT I
PRINCIPLES OF DESCRIPTIVE STATISTICS

CHAPTER 1
NUMBER SENSE

1.1 Identifying Sets of Numbers and Absolute Value

There are a number of different ways to classify numbers. This section will introduce the most common classifications of real numbers so you can explore, apply, and master this concept. Specifically, we will investigate real number classification as it applies to commonly encountered numbers.

Natural Numbers:

The natural numbers start with 1 and include the numbers {1, 2, 3, 4 . . .} and so on. The ellipsis (. . .) symbol indicates that you continue in the pattern established. These are the "counting" numbers and are all positive. Zero is not considered a "natural number."

Whole Numbers:

The whole numbers start with zero and are composed of the numbers {0, 1, 2, 3, 4 . . .}. The whole numbers are all the natural numbers, including zero.

Integers:

The integers are the whole numbers and their opposites: the positive whole numbers, the negative whole numbers, and zero. Integers can be shown as the set {. . . −3, −2, −1, 0, 1, 2, 3 . . .}.

Rational Numbers:

The rational numbers are made up of all integers plus numbers that can be represented as a ratio p/q where p is any integer and q is any non-zero integer. This includes terminating decimals, which end naturally without rounding, and repeating decimals, which have a pattern of numbers that repeats infinitely but does not end. For example, 0.357 is a terminating decimal, and 0.818181818181 . . . is a repeating decimal. The repeating decimal can also be written $0.\overline{81}$ to show that the sequence repeats itself.

Irrational Numbers:

An irrational number is a number with a decimal that neither terminates nor repeats. For example, $\sqrt{2} \approx 1.4142135624\ldots$ is a decimal that does not repeat itself, but that continues infinitely. Any square root of a number that is not a perfect square is an irrational number. The value $\pi \approx 3.1415926536\ldots$ is also classified as an irrational number. We often approximate these irrational numbers with terminating decimals. These approximations, like $\pi \approx 3.14$, are so commonly used that we often think of them as being "equal to" the value of the irrational number, but they are not.

3

Real Numbers:

The combined set of rational and irrational numbers together are called the real numbers. This is the set represented by a "number line."

Here is a graph that summarizes the set of real numbers:

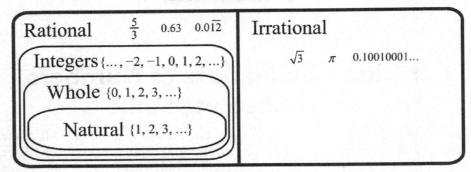

EXAMPLE 1

Classify the numbers 3, $\sqrt{5}$, $\frac{2}{3}$:

	Natural Number	Whole Number	Integer	Rational Number	Irrational Number	Real Number
3	X	X	X	X		X
$\sqrt{5}$					X	X
$\frac{2}{3}$				X		X

- 3 is an integer as well as a rational number ($\frac{3}{1}$ is the fractional equivalent) and a real number— since all rational numbers are real numbers.
- $\sqrt{5}$ is an irrational number since it is the square root and 5 is not a perfect square. It is also a real number since irrationals are real numbers.
- $\frac{2}{3}$ is a rational number because it is a ratio or fraction of two whole numbers and can also be written as a repeating decimal $0.\overline{6}$. It is also a real number.

Number Lines:

Number lines are a visual way of representing all real numbers. An understanding of number lines is necessary to see the relationships in inequality statements as they relate to probability.

There are a few basic ideas underpinning an understanding of number line order:

1. Negative numbers are to the left of zero and positive numbers are to the right of zero.

2. The number line extends infinitely to both the left and right.

3. As you go from left to right, the numbers get larger or increase. This may be confusing with negative numbers because it means that -1 is larger than -2 for example.

Here is a blank number line:

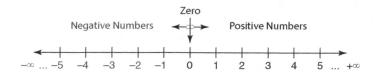

Keeping these ideas in mind, placing numbers correctly on the number line requires that we be able to compare the numbers and determine their relative order from smallest to largest. Perhaps the easiest way to do this is to express all the numbers in decimal form, being sure that each comparison pair has the same number of decimal places. Then we can place the numbers on the number line.

EXAMPLE 2

Place the numbers in the following data set on the number line:

$$-2.4, 0.367, \frac{4}{5}, 0.1, 1.6, \frac{-2}{3}$$

Step 1: Express the numbers in decimal form with the same number of decimal places. In this case, each number should be shown with three decimal places since there is one number in the list with three decimal places.

Using a calculator to find the equivalents when needed:

$$-2.4 = -2.400$$
$$\frac{4}{5} = 0.800$$
$$0.1 = 0.100$$
$$1.6 = 1.600$$
$$-\frac{2}{3} = -0.667 \text{ (rounded)}$$

Step 2: Sort the numbers from smallest to largest: $-2.400, -0.667, 0.100, 0.367, 0.800, 1.600$

Step 3: Now place these numbers on the number line:

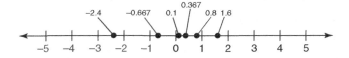

We can also state explicitly that the following inequality is true:

$$-2.4 < -0.667 < 0.1 < 0.367 < 0.8 < 1.6$$

EXAMPLE 3

Graph 3, $\frac{2}{3}$, and $\sqrt{5}$ on a number line:

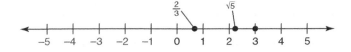

Absolute Value:

On a number line, the distance from zero is called the absolute value. All real numbers have an absolute value that can be determined by examining how many units from zero a number is. For example, the number 3 is three units to the right of zero, while the number -3 is three units to the left of zero. The numbers 3 and -3 have the same absolute value because they are the same distance from zero.

1.1 PRACTICE PROBLEMS

1. Classify each number by checking all of the appropriate boxes:

	Natural Number	Whole Number	Integer	Rational Number	Irrational Number	Real Number
-2						
$\frac{7}{3}$						
$\sqrt{4}$						
5.3						
0						
$\sqrt{7}$						
24						

2. Graph the following numbers on the number line provided: $1.3, -4, \sqrt{5}, \frac{3}{4}$.

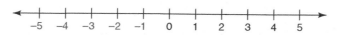

3. Juliette measures the temperature inside a barn at four hour intervals for a day. This is the data set she obtains (in degrees Fahrenheit): three below zero, zero, five, four, one, and one below zero.

   a. List these readings numerically:

   b. Graph these readings on the number line below:

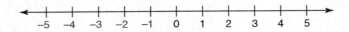

4. What number is the "opposite" of the number 2?

5. Graph both the number 2 and its opposite on the number line below:

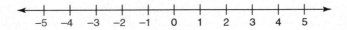

6. What is the absolute value of −5?

7. Why do −10 and 10 have the same absolute value?

1.2 Converting between Fraction, Decimal, and Percentage

It is often necessary in statistics to convert between a fraction, decimal, and percentage. A review of these topics should make it possible for you to convert easily from one form to another.

Fractions:

The basic idea is that a fraction represents "parts of a whole." For example: $\frac{1}{4}$ represents one part out of a total of four parts:

The bottom number in a fraction (the denominator) defines how many parts the whole is split into. The upper number in the fraction (the numerator) tells us the number of parts we are representing in the fraction. In the picture above, we split the whole rectangle into 4 equal parts but only 1 of those parts is shaded, so the fraction can be written as $\frac{1}{4}$.

Decimals:

To convert a fraction to a decimal, we must remember that a fraction also can be interpreted as the numerator divided by the denominator. Thus: $\frac{1}{4} = 1 \div 4 = 0.25$. This is an example of a terminating decimal because it ends without rounding.

We can perform division to change any fraction into a decimal. But not all fractions will result in terminating decimals. For example, $\frac{2}{3}$ results in a repeating decimal: $0.6666\overline{6}$ where the bar over the last 6 indicates that the 6 would repeat indefinitely. We typically will use rounding rules to show this number with only a few decimal places. However, as soon as we round the value, we have chosen to give an approximate answer which is not exactly equal to the original fraction.

Converting a Decimal to a Fraction:

To perform this task, we must know the place value for the decimal place furthest to the right (the last place in the number). Thus: 0.569 has a 9 in the thousandths place so its fractional equivalent is $\frac{569}{1000}$. Place values for several places are shown below:

Percentage Conversions:

To convert a fraction to a percentage, we must first find the decimal equivalent. Then we multiply by 100, which is the same as moving the decimal two places to the right, and add the percent sign. Therefore: $\frac{3}{4} = 0.75 = 75\%$

To convert from a percentage to a fraction, we must first get the decimal equivalent. Then we divide by 100, which is the same as moving the decimal two places to the left, and remove the percent sign. Therefore: $67.2\% = 0.672 = \frac{672}{1000}$.

EXAMPLE 1

Convert $\frac{17}{200}$ from a fraction to a decimal to a percentage:

Step 1: Convert to a decimal by dividing the fraction:

$$\frac{17}{200} = 17 \div 200 = 0.085$$

Step 2: Convert to a percentage by multiplying by 100:

$$0.085 = 0.085 \times 100 = 8.5\%$$

EXAMPLE 2

Convert 38.4% from a percentage to decimal and then fraction:

$$38.4\% = 38.4 \div 100 = 0.384$$

The four is in the thousandths place so the unreduced fraction is $0.384 = \frac{384}{1000}$.

1.2 PRACTICE PROBLEMS

1. Fill in the following table by making all the necessary conversions.

Fraction (leave as an improper fraction if needed)	Decimal (round to four places where needed)	Percentage
$\dfrac{3}{5}$		
	0.589	
		56.2%
	1.36	
$\dfrac{6}{17}$		
		235%
		2.8%
	0.2987	

2. On one committee, four of the seven members are women.

   a. What fraction of the committee is made up of women?

b. What decimal part of the committee is made up of women?

c. What percentage of the committee is made up of women?

3. The newspaper report says that one-third of all voters in the state are Democrats.

   a. What decimal part of the voters are Democrats?

   b. What percent of the voters in the state are Democrats?

1.3 Calculating Proportions from Survey Data

In addition to the ability to convert between fractions, decimals, and percentage, it becomes necessary to be able to analyze the relationships one step further. Consider the basic idea of a fraction as the number of parts (numerator) divided by the total number of parts in the whole (denominator).

Proportion Symbols:

In statistics, we give each of these concepts a symbol.

- Proportion is a fraction symbolized by p
- Total number in a sample is symbolized by n
- Number of parts of the whole we are interested in is usually those who gave a specific answer to a survey, symbolized by x

Now the fraction is a proportion, p, which has a numerator, x, and a denominator, n.

To put this into a formula, we say

$$p = \frac{x}{n}$$

EXAMPLE 1

What proportion of the survey said they would prefer an evening class if 10 out of 40 surveyed gave that response?

> **Step 1:** Identify the variables. In this case the total number in the survey, n, was 40. The number who said they prefer an evening class, x, is 10.

> **Step 2:** Use the formula above to find the fraction or proportion, p.

$$p = \frac{x}{n} = \frac{10}{40} = \frac{1}{4}$$

Solving for x

We can also algebraically rearrange this formula to solve for x.

$$x = p \cdot n$$

This formula allows us to ask questions like the following: How many people answered "yes" if 30% of the 150 people taking the survey answered "yes?" Notice that we first need to convert the percent into a decimal because the proportion can only be a fraction or a decimal number.

$$30\% = 0.30$$

Now, we use the second formula:

$$x = 0.30 \cdot 150 = 45$$

So we know that 45 people said "yes" in the survey.

EXAMPLE 2

What proportion of people said they like to sleep on flannel sheets in the winter if 150 people gave this response out in a survey of 230 people?

> **Step 1:** Read the problem and identify the variables: $n = 230$ *and* $x = 150$.
>
> **Step 2:** Use the formula for proportion: $p = \frac{x}{n} = \frac{150}{230} = .652$

Notice that proportion can be expressed as a fraction or a decimal. But, in this case, the fraction is an exact answer while the decimal has been rounded to three decimal places.

EXAMPLE 3

How many people said that they would vote Democratic if 15% of the 1300 people surveyed gave that response?

> Reading the problem and identifying the variables: $p = 15\% = .15$ *and* $n = 1300$.
>
> Using the second formula for solving for x: $x = p \cdot n = .15 \cdot 1300 = 195$
>
> So, 195 people out of the survey of 1300 said that they would vote Democratic.

1.3 PRACTICE PROBLEMS

1. What is $\frac{3}{4}$ of 1200?

2. If 35 people out of 150 said that they like pancakes, answer the following.

   a. What fraction gave this response?

   b. What proportion?

3. Fifty-two counties in Tennessee are considered part of the Appalachian region, but only 5.8% of these counties produce coal.

   a. What is 5.8% of 52?

   b. Can this <u>exact</u> answer be the number of counties which produce coal? Why or why not?

   c. What is the number of counties that produce coal?

4. There are 55 Appalachian counties in West Virginia and $\frac{5}{11}$ of the counties produce coal. What is the number of counties that produce coal in Appalachian West Virginia?

5. What proportion of a herd of 1500 goats had black spots if an inspector counted 820 goats with black spots?

6. How many dogs are in a shelter if 55% of the 200 animals are dogs?

1.4 Calculating Relative Frequency as Fraction, Decimal, and Percentage

It is often necessary to examine a frequency table and be able to convert the frequency count in each class into a fraction of the total, a decimal part of the total or percentage.

EXAMPLE 1

Consider the frequency table given below:

Class	Frequency Count
0–9	42
10–19	58
20–29	50

Step 1: To find the fractional part represented by each class in the table, we need to find the total frequency count:

$$42 + 58 + 50 = 150$$

Step 2: The fraction represented by each class is the number in the class, divided by the total count.

Class	Frequency Count	Fraction
0–9	42	$\dfrac{42}{150}$
10–19	58	$\dfrac{58}{150}$
20–29	50	$\dfrac{50}{150}$

Step 3: The decimal part is simply converting this fraction into a decimal.

Class	Frequency Count	Fraction	Decimal
0–9	42	$\dfrac{42}{150}$	0.2800
10–19	58	$\dfrac{58}{150}$	0.3867
20–29	50	$\dfrac{50}{150}$	0.3333

(Note that the decimal equivalent has been rounded to the fourth place.)

Step 4: Finally, the percent for each class is found by converting the decimal part into a percent.

Class	Frequency Count	Fraction	Decimal	Percentage
0–9	42	$\dfrac{42}{150}$	0.2800	28%
10–19	58	$\dfrac{58}{150}$	0.3867	38.67%
20–29	50	$\dfrac{50}{150}$	0.3333	33.33%

1.4 PRACTICE PROBLEMS

1. Use your knowledge of conversions between fractions, decimals, and percent, to fill in the frequency tables in the practice problems.

Class	Frequency Count	Fraction	Decimal	Percentage
3–8	2			
9–14	12			
15–20	64			
21–26	18			
27–32	7			

2. The following table shows the frequency counts for 25 students on a statistics exam. Complete the conversions for the remainder of the table.

Grade Range	Count	Fraction	Decimal	Percentage
30–39	1			
40–49	1			
50–59	2			
60–69	7			
70–79	8			
80–89	4			
90–100	2			

3. Complete the table below.

Age Range	Count	Relative Frequency (as a fraction)	Relative Frequency (as a decimal)	Relative Frequency (as a percentage)
1–3	8			
4–6	6			
7–9	12			
10–12	11			
13–15	9			

4. Consider the frequency distribution for the percent of high school completion for the counties in Appalachia.

Class	Frequency
50.0–59.9	1
60.0–69.9	22
70.0–79.9	166
80.0–89.9	211
90.0–99.9	20

a. How many total counties are there are there in Appalachia?

b. Estimate the percent of high school completion that is "average" for a county in Appalachia.

c. Use the table above, convert it to a relative frequency table and record the results in the table below.

Class	Relative Frequency (decimal)	Relative Frequency (percentage)
50.0–59.9		
60.0–69.9		
70.0–79.9		
80.0–89.9		
90.0–99.9		

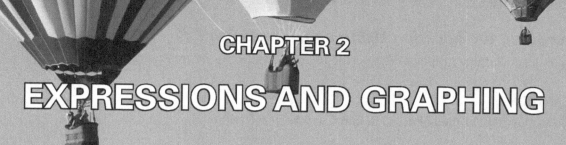

CHAPTER 2
EXPRESSIONS AND GRAPHING

2.1 Plotting Points in the Cartesian Plane

To describe the location of points on a plane, we use a rectangular coordinate system which is called the Cartesian plane. This system is simply two number lines set perpendicular to each other with an intersection at the zero, called the *origin*, of each number line.

Cartesian plane:

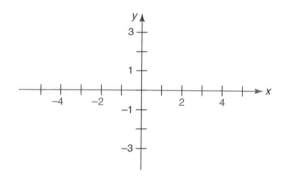

Plotting the Location of a Given Ordered Pair

Locations on the plane are represented by ordered pairs of numbers (x, y) where x indicates the position along the x-axis and y indicates the position along the y-axis. To plot points, begin at the center or origin for both x-and y-values.

EXAMPLE 1

Plot the location of the ordered pair (3, −2):

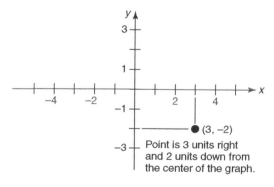

Point is 3 units right and 2 units down from the center of the graph.

21

Describing the Location Using an Ordered Pair

We can also examine graphs and describe the location using ordered pairs (x, y).

EXAMPLE 2

State the location of the points A, B, and C on the graph below.

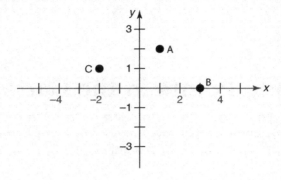

- Point A appears to be 1 unit right on the x-axis and 2 units up on the y-axis from the center of the graph $(0, 0)$. Therefore, the ordered pair is $(1, 2)$.
- Point B appears to be 3 units right and 0 units up from the center of the graph. Therefore, the ordered pair is $(3, 0)$.
- Point C appears to be 2 units left and 1 unit up from the center of the graph. Therefore, the ordered pair is $(-2, 1)$.

2.1 PRACTICE PROBLEMS

1. Identify the location of the labeled points:

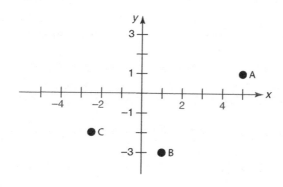

   a. Point A:

   b. Point B:

   c. Point C:

2. Plot and label the following ordered pairs on the Cartesian graph below:

   Point A: $(0, -3)$ Point B: $(1, 2)$ Point C: $(-1, -1)$ Point D: $(-4, 2)$

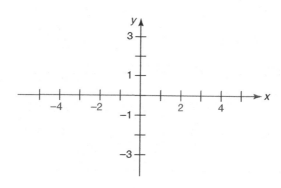

2.2 Introduction to Statistical Symbolic Notation

Statistical Symbols

Symbols are commonly used in math classes as shorthand versions of words or phrases. This is very useful since the concepts "named" by the symbols are often used in other equations and formulas. In statistics, a number of the symbols are taken from the Greek, rather than the English, alphabet. Therefore, it may be challenging to learn to recognize the symbols, know how to write them yourself, and understand what they represent. However, it is necessary to acquire the vocabulary of statistics in order to do the work required. The table below contains the most commonly used symbols, their names, if they are from the Greek alphabet, and what they represent.

Symbol	"Name"	What is represented
$\bar{x}$	x-bar	Mean of a sample
μ	Lower case, Greek letter "mu"	Mean of a population
$\tilde{x}$	x-tilde	Median
s		Standard Deviation of a sample
σ	Lower case, Greek letter "sigma"	Standard Deviation of a population
s^2	"s squared"	Variance of a sample
σ^2	"sigma squared"	Variance of a population
$\hat{p}$	p-hat	Sample proportion
n		The total number in the data set (sample)
N		The total number in the population
x		This is the variable name used to represent any data value.
Σ	Upper case, Greek letter "sigma"	Summation symbol—this is different, it is an operation!! So Σx *means to add up all the values of x.*

Notice that more than one symbol is used for mean, standard deviation, and variance. This is because it is important to specify whether you are dealing with data that constitutes a sample or the entire population. In all cases, using the correct symbol is critical. You are not allowed to make up your own symbols or blur with the distinction between sample and population.

EXAMPLE 1

$\bar{x} = \frac{\Sigma x}{n}$ **should be interpreted like this:**

The mean of a sample is equal to the sum of all the data values divided by the total number of data values.

EXAMPLE 2

$\mu = \frac{\sum x}{N}$ **should be interpreted like this:**

The mean of the population is equal to the sum of all values in the population divided by the total number in the population.

2.2 PRACTICE PROBLEMS

1. Match the following symbols and statistical name.

   n A. Sample Mean
   μ B. Sample Standard Deviation
   s C. Population Mean
   σ^2 D. Population Standard Deviation
   $\bar{x}$ E. Sample Variance
   σ F. Population Variance
   s^2 G. Sample Size

2. In a survey of randomly selected 180 college professors, the mean amount of time spent grading papers each week was 4.5 hours. What is the correct symbol for this mean?

3. A small country has 600,000 citizens. Researchers select a group of 2,400 people to complete a survey on happiness. The results revealed that 95% reported being "very happy with their life." Fill in the following.

   $N =$

   $n =$

   $\hat{p} =$

4. The US Census reports that the mean household size is 2.58 people with a standard deviation of 1.8 people. Based on this information, identify the following:

   $\mu =$

   $\sigma =$

5. How could you explain the difference between the symbols used in describing the mean in questions 2 and 4?

2.3 Order of Operations in Measures of Central Tendency

Here is a basic review of the order of operations.

Order of Operations Review

You will frequently use the order of operations to perform algebraic computations. The order of operations is as follows:

Parentheses | Exponents | Multiplication | Division | Addition | Subtraction

1. Perform the operations inside parentheses and other grouping symbols such as brackets and division bars first.
2. Then exponents.
3. Then multiplication and division, from left to right.
4. Then addition and subtraction, from left to right.

The acronym **PEMDAS** is used as a mnemonic device to help remember the order of operations.

Let's review the most basic rules of arithmetic operations.

Integer Addition:

Positive + Positive = Positive: $4 + 7 = 11$

Negative + Negative = Negative: $(-5) + (-2) = -7$

When you have a problem with a positive value and a negative value, you need to subtract the absolute values of the two numbers and then the answer gets the sign of the number with the larger absolute value:

$$-11 + 2 = -9$$

Integer Subtraction:

Negative − Positive = Negative: $-7 - 6 = -7 + (-6) = -13$

Positive − Negative = Positive + Positive = Positive: $7 - (-2) = 7 + 2 = 9$

Negative − Negative = Negative + Positive: Therefore, you need to consider the absolute values and proceed as for addition: $-2 - (-8) = -2 + 8 = 6$

Integer Multiplication:

Positive × Positive = Positive: $8 \cdot 2 = 16$

Negative × Negative = Positive: $-4 \cdot (-3) = 12$

Negative × Positive = Positive × Negative = Negative: $-8 \cdot 5 = 5 \cdot (-8) = -40$

Integer Division:

Positive ÷ Positive = Positive: $\frac{6}{2} = 3$

Negative ÷ Negative = Positive: $\frac{-40}{-5} = 8$

Negative ÷ Positive = Positive Negative = Negative: $\frac{15}{-3} = \frac{-15}{3} = -5$

EXAMPLE 1
Evaluate 6 · (2 + 8).

> **Step 1:** We note that the order of operations requires us to solve inside the parenthesis.

$$6 \cdot (2 + 8) = 6 \cdot 10$$

> **Step 2:** We multiply this result,

$$6 \cdot 10 = 60$$

EXAMPLE 2
Evaluate $\mu + 2\sigma$ when the mean is 6.3 and the standard deviation is 1.2.

> **Step 1:** Substitute $\mu = 6.3$ *and* $\sigma = 1.2$.

$$6.3 + 2 \cdot 1.2$$

> **Step 2:** Use the order of operations to simplify. The multiplication must be done first.

$$6.3 + (2 \cdot 1.2) = 6.3 + 2.4$$

> **Step 3:** Then you would add,

$$6.3 + 2.4 = 8.7$$

EXAMPLE 3
Evaluate $(3 + 5)^2 - 4 \cdot 6$.

> **Step 1:** The parenthesis must be done first.

$$(3 + 5)^2 - 4 \cdot 6$$
$$= 8^2 - 4 \cdot 6$$

> **Step 2:** The exponent must be done next.

$$8^2 - 4 \cdot 6$$
$$= 64 - 4 \cdot 6$$

> **Step 3:** Now we must do the multiplication.

$$64 - 4 \cdot 6$$
$$= 64 - 24$$

> **Step 4:** Finally, subtract the two values.

$$64 - 24 = 40$$

Order of Operations and the Mean of a Sample:

A common formula to consider is the formula for the mean of a sample

$$\bar{x} = \frac{\sum x}{n}$$

Because of the division bar, the numerator operation should be done first. This is like putting the numerator in a parenthesis:

$$\overline{x} = \frac{(\Sigma\, x)}{n}$$

In this formula, you must also remember that the symbol Σ means that you must do a summation. Since the entire phrase is, Σx the directive is to sum all the values of x in the sample. Then you would finish by dividing by n, the total number in the data set.

EXAMPLE 4

Consider the calculation of the mean for the data set given below:

2, 6, 2, 4, 5

Step 1: Determine how many numbers are in the data set. Since there are five numbers in the data set, $n = 5$.

Step 2: Substituting into the formula for the mean:

$$\overline{x} = \frac{\Sigma x}{n} = \frac{2+6+2+4+5}{5} = \frac{19}{5} = 3.8$$

- Note that the summation was done first as required by Step 1 of the order of operations. Step 2 was skipped because there are no exponents. Finally, the division of the numerator by the denominator satisfies Step 3 of the order of operations.

Order of Operations and the Midrange:

Specifically, we need to use the order of operations correctly when applied to the midrange which is another, although less commonly used, measure of center. The formula for computing the midrange is as follows:

$$midrange = \frac{Maximum + Minimum}{2}$$

As noted earlier, the division bar means the numerator of the fraction is implied to be in a parenthesis. Thus:

$$midrange = \frac{(Maximum + Minimum)}{2}$$

EXAMPLE 5

Consider finding the midrange of the following data list:

5, 8, 4, 6, 14, 40, 6, 8, 10, 9

Step 1: We need to determine the maximum number and minimum number in the data set. We do this by sorting the data list from lowest to highest.

When sorted this data list is: 4, 5, 6, 6, 8, 8, 9, 10, 14, 40.

So the maximum number is 40 and minimum is 4. Substituting into the midrange equation:

$$midrange = \frac{(Maximum + Minimum)}{2} = \frac{(40 + 4)}{2}$$

Step 2: Applying the order of operations, do the operation in the numerator.

$$midrange = \frac{(40 + 4)}{2} = \frac{44}{2}$$

Step 3: There are no exponents, so we next do the division.

$$midrange = \frac{(Maximum + Minimum)}{2} = \frac{(40 + 4)}{2} = \frac{44}{2} = 22$$

2.3 PRACTICE PROBLEMS

1. Use order of operations to simplify:

$$12 + 2 \cdot 5$$

2. Use order of operations to simplify:

$$5^2 + 7 \cdot 9 \div 3 - 2$$

3. Use order of operations to simplify:

$$\frac{20 - 12}{4}$$

4. Use order of operations to simplify:

$$\frac{7 + 4 + 9 + 2 + 3}{5}$$

5. Evaluate

$$\overline{x} = \frac{\sum x}{n}$$

for x-values of 7, 8, 9, 10 and n = 4.

6. Evaluate

$$midrange = \frac{(Maximum + Minimum)}{2}$$

for the following set of data 3, 5, 7, 18, 22, 23.

2.4 Evaluating Rational Roots in Measures of Variation

Radicals

In evaluating the standard deviation by hand, it is necessary to take a square root. The $\sqrt{}$ symbol is called the radical symbol. A square root is written as $\sqrt[2]{}$. In this case, the index number (which tells you which root you are seeking) is the number 2. Generally, however, we do not write the index number 2, instead assume that $\sqrt{} = \sqrt[2]{}$. A square root is also a fractional exponent with the index number being the denominator of the fractional exponent. Thus, $\sqrt{x} = x^{\frac{1}{2}}$.

(The rest of this section contains discussion of rational roots beyond the square root. These concepts, while important, are not vital to the study of statistics. Examples with an asterisk may be safely omitted.)

However, if the index number is not a 2, then we must write it. For example, the cube (third) root is written as $\sqrt[3]{}$. The cube root is also a fractional exponent, since $\sqrt[3]{x} = x^{\frac{1}{3}}$. Notice that the index number is <u>always</u> the denominator of the fractional root.

When you use the radical symbol, you are asking yourself, "What number would I have to multiply by itself the index number of times to get the value under the radical symbol?"

EXAMPLE 1

Find $\sqrt[2]{36}$.

> To determine the answer, we need to ask what number could be multiplied by itself 2 times and get 36? The answer is 6, because $6 \cdot 6 = 36$.

*EXAMPLE 2

Find $\sqrt[3]{125}$.

> To determine the answer, we need to ask what number could be multiplied by itself 3 times and get 125? The answer is 5, because $5 \cdot 5 \cdot 5 = 125$.

Of course, most roots do not turn out to be perfect whole numbers. This is why we use the calculator to do the evaluations for us and give us decimal approximations of the value of the root.

EXAMPLE 3

Use the calculator to find the square root of 18.

$$\sqrt{18} \sim 4.24.$$

> This means that $4.24 \cdot 4.24$ *is about* 18. Remember that we rounded the answer of 4.24, so it is not an exact square root.

*EXAMPLE 4

Use the calculator to find $\sqrt[4]{200}$.

There is not a button on the calculator specifically for the fourth root. So, we use a more general procedure:

Step 1: Type in the index number of 4.

Step 2: Press the MATH button and go down to the function number 5, $\sqrt[n]{}$ and press ENTER.

Step 3: Now the window should show the calculation you want: $\sqrt[4]{}$ so all you need to do is type in the number 200 and press ENTER.

$$\sqrt[4]{200} = 3.76 \text{ rounded to the hundredth place}$$

The formula for standard deviation is a more complex case of using the order of operations because it does contain an exponent (the square root). Let's look at the formula and break it down:

$$s = \sqrt{\frac{\sum (x - \bar{x})^2}{n-1}} = \left(\frac{\sum (x - \bar{x})^2}{n-1}\right)^{1/2}$$

Remember that the square root is the same as raising the stuff under the square root symbol (the radicand) to the $\frac{1}{2}$ power.

So, applying the order of operations, Step 1 means that we must evaluate the expression

$$\frac{\sum (x - \bar{x})^2}{n-1}$$

before we can take the square root.

Considering that the expression $\frac{\sum (x-\bar{x})^2}{n-1}$ is a fraction, we know that we must do the operations in the numerator first. That is, we must evaluate $\sum (x - \bar{x})^2$ before we do anything else, but note that there is a parenthesis here also. So Step 1 of the order of operations forces us to find all the values of $(x - \bar{x})^2$ and then add them all together. Once we have done that summation, then we can divide by $n-1$. Finally, we will take the square root of the quotient, the answer obtained by the division of numerator by denominator.

Complicated, huh? An example using real numbers should help.

EXAMPLE 5
Consider the data set {2, 6, 2, 4, 5} and calculate the standard deviation.

Tabular form is being used because it makes the process easier.

Step 1: Determine the mean. Using the data set {2, 6, 2, 4, 5}, the mean $\bar{x}$ is 3.8.

Step 2: Subtract the mean from each data value then square the result.

x	$(x - \bar{x})$	$(x - \bar{x})^2$
2	$(2 - 3.8) = -1.8$	$(-1.8)^2 = 3.24$
6	$(6 - 3.8) = 2.2$	$(2.2)^2 = 4.84$
2	$(2 - 3.8) = -1.8$	$(-1.8)^2 = 3.24$
4	$(4 - 3.8) = 0.2$	$(0.2)^2 = 0.04$
5	$(5 - 3.8) = 1.2$	$(1.2)^2 = 1.44$

Step 3: Now we need to perform the operation in the numerator by finding the sum of those values in the last column:

$\sum (x - \bar{x})^2$	$3.24 + 4.84 + 3.24 + .04 + 1.44 = 12.8$

Step 4: Next, perform the division under the radical sign or inside the parentheses.

$$s = \left(\frac{\sum (x - \bar{x})^2}{n-1} \right)^{1/2} = \left(\frac{12.8}{(n-1)} \right)^{1/2} = \left(\frac{12.8}{(5-1)} \right)^{1/2} = \left(\frac{12.8}{4} \right)^{1/2}$$

Then, because the square root is equivalent to raising the amount inside the square root sign to the one-half power, this is the same as saying:

$$s = \left(\frac{12.8}{4} \right)^{1/2} = (3.2)^{1/2}$$

$$= \sqrt{3.2} = 1.789$$

(rounded to three decimal places)

2.4 PRACTICE PROBLEMS

1. Find the roots indicated:

   a. $\sqrt{16}$

   b. $\sqrt{24}$

   c. $\sqrt[3]{125}$

   d. $\sqrt[4]{16}$

2. Express the root shown as a fractional root and solve.

   a. $\sqrt{5}$

   b. $\sqrt[5]{32}$

3. Find the value of the standard deviation using the order of operations. Show all intermediate steps.

$$s = \sqrt{\frac{6(26870) - (324)^2}{6(6-1)}}$$

4. Find the value of the standard deviation. Show all intermediate steps.

$$s = \sqrt{\frac{6(105965) - (795)^2}{6(6-1)}}$$

2.5 Evaluating Expressions in Measures of Relative Standing

One crucial skill needed to be successful in mathematics is the ability to correctly connect the numbers in a word problem with the symbolic concept that they represent. In addition, you need to then be able to substitute properly into the formula given and use the order of operations correctly to solve.

Order of Operations and z-scores:

Of course, there are several nuances in the use of the order of operations.

For example, consider the formula for the -score:

$$z = \frac{(x - \mu)}{\sigma}$$

The parentheses have been added so that operation in the numerator can be done first. Consider the calculation for the z-score using real numbers, rather than symbols:

EXAMPLE 1

Evaluate the formula $z = \frac{x - \mu}{\sigma}$ for $x = 60$, $\mu = 20$, and $\sigma = 15$

> **Step 1:** Substitute the values into the formula. So when we substitute:
>
> $$z = \frac{(x - \mu)}{\sigma} = \frac{(60 - 20)}{15}$$
>
> **Step 2:** We do the numerator operation first:
>
> $$z = \frac{(60 - 20)}{15} = \frac{40}{15}$$
>
> **Step 3:** Now we divide the numerator by the denominator:
>
> $$z = \frac{40}{15} = 2.667$$
>
> (rounded to three decimal places)

EXAMPLE 2

What is the z-score for a height of 70 inches when the mean height is 62.5 inches with a standard deviation of 2.3 inches?

We will be solving for the z-score. To properly match the numerical values with the concepts, we need to analyze the wording. Usually, the value is placed in close proximity to the concept measured. Note that it is critical that you KNOW the symbols for each concept. Here is a highlighted version of the problem to help clarify this idea:

What is the z-score for a height of 70 inches when the mean height is 62.5 inches with a standard deviation of 2.3 inches ?

> **Step 1:** Determine the values from reading the question. Noting the wording and the proximity allows the following conclusions:
>
> $$x = 70 \qquad \mu = 62.5 \qquad \sigma = 2.3$$

Step 2: So, we now substitute into the formula the known values for x, μ, and σ.

$$z = \frac{x - \mu}{\sigma} = \frac{(70 - 62.5)}{2.3}$$

Step 3: Do the operation in the numerator and then divide the numerator by the denominator.

$$z = \frac{x - \mu}{\sigma} = \frac{(70 - 62.5)}{2.3} = \frac{7.5}{2.3} = 3.26$$
(rounded to two decimal places)

Evaluating for the Maximum and Minimum Usual Values

Another place where we need to use the order of operations properly occurs when we use the empirical rule for finding the maximum and minimum usual values. In a typical problem, we are given the mean and standard deviation for a set of data and then asked to determine what data values are considered the maximum and minimum "usual" values. To find those boundary values, we use the formulas below:

$$Maximum\ Usual = \bar{x} + 2 \cdot s$$
$$Minimum\ Usual = \bar{x} - 2 \cdot s$$

EXAMPLE 3

I am told by the Mars Candy Company that standard size bags of M&M candies have a mean of 210 with a standard deviation of 11. Would it be unusual for me to buy a bag and find that there were only 175 candies in it?

Step 1: Substitute into the formula for the minimum usual and maximum usual values.

$$Maximum\ Usual = \bar{x} + 2 \cdot s = 210 + 2 \cdot 11$$
$$Minimum\ Usual = \bar{x} - 2 \cdot s = 210 - 2 \cdot 11$$

Step 2: Follow the order of operations by doing the multiplication first.

$$Maximum\ Usual = 210 + (2 \cdot 11) = 210 + 22$$
$$Minimum\ Usual = 210 - (2 \cdot 11) = 210 - 22$$

Step 3: Do addition or subtraction to complete the calculation.

$$Maximum\ Usual = 210 + 22 = 232$$
$$Minimum\ Usual = 210 - 22 = 188$$

My conclusion is that it would be unusual to get a bag with only 175 candies because this value is below the minimum usual number.

2.5 PRACTICE PROBLEMS

1. Evaluate $z = \frac{x-\mu}{\sigma}$ for $x = 28$, $\mu = 35$, and $\sigma = 2$.

2. What is the z-score associated with the mortality rate (per hundred-thousand) in Knox County, Tennessee relative to all the other counties in Appalachia? The following information is known for this situation:

$$mortality\ rate\ in\ Knox\ County,\ TN = x = 913.9$$
$$mean\ mortality\ rate\ in\ Appalachia = \mu = 1095.82$$
$$standard\ deviation = \sigma = 179.88$$

3. What is the meaning of the z-score from problem #1? Is it good or bad to have a negative z score for the variable of mortality rate?

4. The data from the Bureau of Labor Statistics shows that the mean salary in the United States in 2012 was $54875.67 with a standard deviation of $25256.05. Dentists had a mean salary in the same year of $149310. What z-score is associated with the average salary of dentists?

5. Is the z-score found in problem #3 unusually high or low? Why or why not?

6. When I bought my new car, I paid extra for better tires. The salesman guaranteed that the mean life of these years was 60000 miles with a standard deviation of 7800 miles. I had to replace my tires at 50000 miles. Would 50000 miles be considered an unusually low lifespan for the tires?

7. In a given data set, the mean is at 8.3 with a standard deviation of 1.2. The data is bell-shaped. Is the data value at 13.2 considered usual or unusual? Show your work.

8. In my statistics class I learned that men's heights have a mean at 69.5 inches with a standard deviation of 2.4 inches. Is my height of 78 inches, usual or unusual?

UNIT II

PRINCIPLES OF DISCRETE PROBABILITY DISTRIBUTIONS

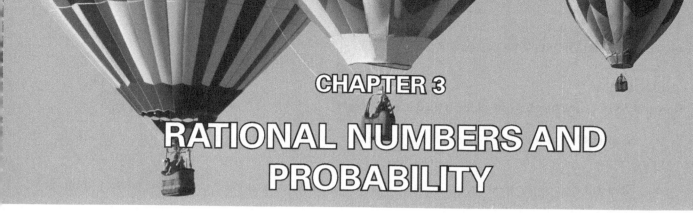

CHAPTER 3
RATIONAL NUMBERS AND PROBABILITY

3.1 Computing Probabilities as Fractions, Decimals, and Percentages

Fractions to Decimals

The first step to convert a probability result in fractional form into decimal notation is to divide the numerator (top number) of the fractional form by the denominator (bottom number) of the fraction. This will give you the decimal form of the fraction and occasionally the answer will be a terminal decimal: a decimal which ends naturally because the division comes out even. In many cases, however, the decimal number does not end naturally or you are asked for an answer which is rounded to a specific number of decimal places.

Place Value

In order to do this, you need to have knowledge of "place value" for each of the digits in the number. The following chart is a "map" of place value with special emphasis on the places to the right of the decimal place.

Place Value Chart
Millions	Hundred thousands	Ten thousands	Thousands	Hundreds	Tens	Ones	• Decimal point	Tenths	Hundredths	Thousandths	Ten thousandths	Hundred thousandths

Notice that there is a "ones" place to the left of the decimal, but there is no "oneth" place to the right!

Process of Rounding Numbers

Step 1: Decide which place value you must round to.

Step 2: Look at the number in the place to the right.

- If the value in this place is less than five (0 to 4), there is no change to the place value.
- If the value in this next place is greater than or equal to five (5 to 9), there is a change of +1 to the place value.

EXAMPLE 1

Round 0.5878 to the hundredths place (two decimal places).

> **Step 1:** The hundredths place is the second place to the right of the decimal. There is an 8 is this place.
>
> **Step 2:** Look at the next place; there is a 7 here. Since the 7 is in the 5 to 9 range, we round the 8 which was in the hundredths place to a 9.
>
> **Answer:** The rounded number is 0.59

Decimals to Fractions

Conversely, we may be asked to express a decimal probability as a fraction. To do this, we also have to understand place value because the denominator of the fraction is determined by the place value of the number furthest right of the decimal. In probabilities, all of the numbers will be between 0 and 1. Therefore, we don't need to worry about numbers to the left of the decimal. We may also need to reduce the fraction but this can easily be done using the calculator.

To write a decimal probability as a fraction, do the following:

> **Step 1:** Identify the place value of the digit furthest to the right.
>
> **Step 2:** Write this number as the denominator (the bottom number) of a fraction.
>
> **Step 3:** The numerator of the fraction is simply the digits to the right of the decimal place.

EXAMPLE 2

Change 0.126 to a fraction.

> **Step 1:** The number furthest to the right is the 6 which is in the thousandths place. So the denominator of the fractional form is 1000.
>
> **Step 2:** The numerator of the fraction is simply the digits to the right of the decimal place: 126.
>
> **Answer:** The full fraction would be $\frac{126}{1000}$ which can be reduced to $\frac{63}{500}$.

EXAMPLE 3

Change the probability of 0.00089 to a fraction.

> **Step 1:** The number furthest to the right is the 9 which is in the hundred-thousandths place. So the fraction has a denominator of 100,000.
>
> **Step 2:** The numerator of the fraction is 89, do not include leading zeros.
>
> **Answer:** The full fraction would be $\frac{89}{100000}$ which will not reduce.

Convert Fractions to Decimals and Percentages

In statistics, you are often required to determine the requested probability as a fraction, decimal, and percentage. The basic procedure used to change a decimal into a percentage is to multiply by 100—which is equivalent to moving the decimal place two places to the right and adding the percent sign (%).

EXAMPLE 4

A survey of your class of 20 reveals that 7 of them wear glasses. If this sample of 20 is used as a random sample of sufficient size to predict the probability that a student in your college wears glasses, then you can state the probability in several ways.

As a fraction, the total number in the survey is the denominator, while the number who wear glasses is the numerator.

$$P(wears\ glasses) = \frac{7}{20}$$

As a decimal, you need to divide the fraction above:

$$P(wears\ glasses) = 7 \div 20 = 0.35$$

As a percentage, multiply the decimal by 100 and add a percent sign:

$$P(wears\ glasses) = 0.35 \cdot 100\% = 35\%$$

Percentages to Decimals

You will also be asked to go the other direction and change a percentage into a decimal number. The basic procedure here is to divide by 100—which is equivalent to moving the decimal two places to the left.

EXAMPLE 5

Convert 23.7% to a decimal.

Divide the number 23.7 by 100:

$$23.7\% = 23.7 \div 100 = 0.237$$

EXAMPLE 6

In a political survey of 215 registered voters, 166 of them said that they would vote for Woodruff if the election were held today.

As a fraction:

$$P(vote\ for\ Woodruff) = \frac{166}{215}$$

As a decimal:

$$P(vote\ for\ Woodruff) = \frac{166}{215} = 166 \div 215 = 0.772$$

As a percentage:

$$P(vote\ for\ Woodruff) = 0.772 \cdot 100\% = 77.2\%$$

3.1 PRACTICE PROBLEMS

1. Convert the fraction $\frac{3}{9}$ into a decimal rounded to the hundredths place.

2. Convert the fraction $\frac{12}{66}$ into a decimal rounded to the third decimal place.

3. Convert the fraction $\frac{35}{126}$ into a decimal rounded to the ten-thousandths place.

4. There are 54 Appalachian counties in Kentucky and 19 of them produce coal. What fraction of the counties in Appalachian Kentucky produces coal?

5. Express the answer to problem 4 as a decimal rounded to two decimal places.

6. In a city in Connecticut, 118 firefighters took a promotion exam, with 41 people taking the exam for promotion to captain. What fraction of the promotion exam takers were taking the captain's test?

7.   What decimal proportion were taking the captain's test? (Round to the thousandths place.)

3.2 Determining Complementary Probabilities: Subtracting Fractions

Complementary Events

To find the probability of the complement of any event, we need to subtract the fractional probability from 1. The mathematical logic for this is shown below:

$$P(event\ A) + P(not\ event\ A) = 1$$

Rearranging this equation using algebra:

$$P(not\ event\ A) = 1 - P(event\ A)$$

For some students this subtraction offers a challenge because they are subtracting a fraction from the whole number 1. When we subtract fractions we need to have a common denominator, which means we need to show the number 1 with the same denominator as the fractional probability. But this is easy if you remember that "any number divided by itself equals one." That is,

$$\frac{4}{4} = 1 = \frac{7}{7} = \frac{0.2}{0.2}$$

Procedure for Subtracting Fractions from 1

Step 1: Change the 1 into a fraction with both the numerator and the denominator having the same number as the denominator of the fractional probability.

Step 2: Combine the arithmetic operation in the numerator over the common denominator.

Step 3: Solve the arithmetic operation in the numerator.

EXAMPLE 1

Let's say we know that the probability of winning some game of chance, *P*(win), is $\frac{3}{53}$. What is the probability of losing (not winning)?

Step 1: Recognize that losing is the complement of winning. So:

$$P(not\ win) = 1 - P(win) = 1 - \frac{3}{53}$$

Step 2: Convert the 1 into a fraction with the denominator of 53.

$$P(not\ win) = \frac{53}{53} - \frac{3}{53}$$

Step 3: Do the subtraction in the numerator.

$$P(not\ win) = \frac{53}{53} - \frac{3}{53} = \frac{53-3}{53} = \frac{50}{53}$$

3.2 PRACTICE PROBLEMS

1. If we know that 41 of 118 candidates were taking the test for promotion to captain, what fraction were taking any other test? That is, what fraction were not taking the captain's test? Show your work.

2. If the fraction of coal-producing counties in Appalachian Tennessee is $\frac{3}{52}$, what is the fraction of counties in Appalachian Tennessee that do not produce coal? Show your work.

3. In a group of 654 youth, 65 of them admitted to smoking. What fraction of the youth smoke? Show your work.

4. What fraction of the youth does not smoke? Show your work.

5. In a standard deck of 52 cards, 4 cards are Aces. If a card is drawn at random from a shuffled deck, what is the probability that the card drawn is not an Ace?

6. A Twister® spinner has 16 colored sectors, with 4 each of red, green, yellow, and blue. In an intense game, you need a blue to win. What is the probability you will lose?

3.3 Using the Addition Rule: Adding and Subtracting Fractions

When we perform operations using the addition rule for probability, we will need to both add and subtract fractions.

Procedure for Adding and Subtracting Fractions

Step 1: Verify that all the denominators are the same. If they are not then we need to find a common denominator.

Step 2: Combine the arithmetic operations in the numerator over the common denominator.

Step 3: Solve the arithmetic operation in the numerator.

EXAMPLE 1

Use the contingency table below to find the probability indicated.

	Male	Female	TOTALS
Fears flying	7	5	12
Does not fear flying	3	20	23
TOTALS	10	25	35

If one person is selected at random from the table above, what is the probability of selecting a female or someone who does not fear flying?

The addition rule application would require us to solve this problem:

$$P(\text{female OR doesn't fear flying}) = P(\text{female}) + P(\text{doesn't fear flying}) - P(\text{female and doesn't fear flying})$$

We would get these individual fractional probabilities:

$$P(\text{female}) = \frac{25}{35}$$

$$P(\text{doesn't fear flying}) = \frac{23}{35}$$

$$P(\text{female and doesn't fear flying}) = \frac{20}{35}$$

Substituting into the addition rule formula:

$$P(\text{female OR doesn't fear flying}) = \frac{25}{35} + \frac{23}{35} - \frac{20}{35} =$$

All of the denominators are the same, so we can combine over the common denominator:

$$P(\text{female OR doesn't fear flying}) = \frac{25}{35} + \frac{23}{35} - \frac{20}{35} = \frac{25 + 23 - 20}{35} = \frac{28}{35}$$

EXAMPLE 2

A class has 23 people in it. Eleven are female and the rest are male. Fifteen people in the class are under 25 years of age and twelve of these are male. The rest of the class is 25 or older. If I select one person at random from the class list, what is the probability of selecting a male OR someone under 25?

The addition rule application would require us to solve this problem:

$$P(\text{male or under 25}) = P(\text{male}) + P(\text{under 25}) - P(\text{male and under 25})$$

We would get these fractional probabilities:

$$P(\text{male}) = \frac{12}{23}, \text{ since 12 people in the class are men}$$

$$P(\text{under 25}) = \frac{15}{23}$$

$$P(\text{male and under 25}) = \frac{12}{23}$$

So, applying the addition rule:

$$P(\text{male OR under 25}) = \frac{12}{23} + \frac{15}{23} - \frac{12}{23}$$

All of the denominators are the same, so we can combine the operations in the numerator over the common denominator:

$$P(\text{male OR under 25}) = \frac{12 + 15 - 12}{23} = \frac{15}{23}$$

3.3 PRACTICE PROBLEMS

1. In a box of 12 cupcakes, 4 are red velvet, 6 are salted caramel, and 2 are chocolate. If a cupcake is selected at random, the probability of choosing a red velvet or chocolate cupcake is:

$$P(red\ velvet\ OR\ chocolate) = \frac{4}{12} + \frac{2}{12}$$

Add the fractions to determine the total probability.

2. A coin and six-sided die are tossed at the same time. The probability of tossing either a 2 on the die or a Heads on the coin is given by the addition rule as:

$$P(2\ OR\ Heads) = \frac{1}{6} + \frac{1}{2}$$

Find the overall probability by completing the addition.

3. Consider the contingency table below:

		Gender	
		Female	Male
Smoking Status	Smoker	39	26
	Nonsmoker	279	310

Find the probability of selecting a female or a nonsmoker if you select one name at random.

This probability is shown by the equation below:

$$P(female\ OR\ nonsmoker) = P(female) + P(nonsmoker) - P(female\ nonsmoker)$$

$$= \frac{318}{654} + \frac{589}{654} - \frac{279}{654}$$

Show your work to complete this problem.

4. Consider the contingency table below of counties in Appalachia by state and production of coal.

	AL	GA	KY	MD	SC	TN
Coal	10	0	19	2	0	3
No Coal	27	37	35	1	6	49
TOTAL	37	37	54	3	6	52

Find the probability of selecting a coal producing county or a county in Tennessee if you select one county at random. This probability is shown by the equation below:

$$P(coal\ producing\ or\ TN\ county) = P(coal\ producing) + P(TN\ county) -$$
$$P(coal\ producing\ county\ in\ TN) = \frac{34}{189} + \frac{52}{189} - \frac{3}{189}$$

Show your work to complete this problem.

5. Consider the contingency table below:

	Lieutenant candidate	Captain candidate
White	43	25
Black	19	8
Hispanic	15	8

Find the probability of selecting a Lieutenant candidate or a Hispanic if you select one person at random. This probability is shown by the equation below:

$$P(Lieutenant\ OR\ Hispanic) =$$

$$P(Lieutenant) + P(Hispanic) -$$

$$P(Lieutenant\ candidate\ who\ is\ Hispanic) = \frac{77}{118} + \frac{23}{118} - \frac{15}{118}$$

Show your work to complete this problem.

3.4 Using the Multiplication Rule: Multiplying Fractions

Multiplying fractions differs from the addition and subtraction of fractions because a common denominator is not needed.

Multiplying Fractions

Step 1: Multiply the numerators (top numbers) together. This is the new numerator.

Step 2: Multiply the denominators (bottom numbers) together. This is the new denominator.

EXAMPLE 1

Multiply: $\dfrac{4}{7} \cdot \dfrac{5}{12}$

Step 1: Since $4 \cdot 5 = 20$, the new numerator is 20.

Step 2: Since $7 \cdot 12 = 84$, the new denominator is 84.

Step 3: The unreduced answer is $\frac{20}{84}$.

When using the multiplication rule, more than two fractions may need to be multiplied to find the final probability.

EXAMPLE 2

What is the probability of dealing four consecutive Kings when dealing four cards?

This probability is given below:

$$P(four\ Kings) = \frac{4}{52} \cdot \frac{3}{51} \cdot \frac{2}{50} \cdot \frac{1}{49}$$

Step 1: The numerator is the multiplication $4 \cdot 3 \cdot 2 \cdot 1 = 24$.

Step 2: The denominator is the multiplication $52 \cdot 51 \cdot 50 \cdot 49 = 6497400$.

Step 3: The unreduced answer is $\frac{24}{6497400}$. However, we can reduce this fraction but the calculator is no help here because of the size of the denominator. So, we will have to reduce the old-fashioned way by determining common factors.

Notice that both numbers are even, so both are evenly divisible by 2. Thus,

$$24 \div 2 = 12 \text{ and } 6497400 \div 2 = 3248700$$

Dividing both the numerator and denominator by 2 gives us:

$$\frac{24}{6497400} = \frac{12}{3248700}$$

But, again, both of these numbers are divisible by two:

$$\frac{12}{3248700} = \frac{6}{1624350} = \frac{3}{812175}$$

The only factor of 3 is 3 itself, so we need to check to see if 812175 is evenly divisible by 3, which it is. So, completing our reduction we will divide the numerator and denominator by 3:

$$\frac{3}{812175} = \frac{1}{270725}$$

3.4 PRACTICE PROBLEMS

1. Multiply the following fractions.

   a. $\dfrac{24}{66} \cdot \dfrac{2}{3} =$

   b. $\left(\dfrac{11}{59}\right)\left(\dfrac{6}{7}\right) =$

   c. $\dfrac{25}{111} \cdot \dfrac{12}{23} \cdot \dfrac{2}{5} =$

2. Consider the contingency table given below for several counties in Appalachia by state and coal production:

	AL	GA	KY	SC	TN	VA	WV
Coal	10	0	19	0	3	6	25
No Coal	27	37	35	6	49	19	30
TOTAL	37	37	54	6	52	25	55

The grand total for the table is 266.

If you select two counties randomly from this table, what is the probability you select two counties from Georgia? Assuming you sample without replacement, the probability is given by the application of the multiplication rule below:

$$P(GA \ county \ and \ then \ GA \ county) = P(GA) \cdot P(GA|GA) = \frac{37}{266} \cdot \frac{36}{265}$$

3. Consider the contingency table given below for gender versus smoking status:

		Gender	
		Female	Male
Smoking Status	Smoker	39	26
	Nonsmoker	279	310

The grand total for the table is 654.

If you select two people randomly from this table, what is the probability you select two males? Assuming you sample without replacement, the probability is given by the application of the multiplication rule below:

$$P(male \ and \ then \ male) = P(male) \cdot P(male|male) = \left(\frac{336}{654}\right)\left(\frac{335}{653}\right)$$

3.5 Interpreting Scientific Notation

Scientific notation was developed to allow us to write and perform operations with very large or very small numbers more easily.

A number is in scientific notation when it is broken up as the product of two parts: a number, a, which is greater than or equal to 1 and less than 10, multiplied by 10 raised to an integer power, n. Symbolically, this format is $a \cdot 10^n$.

Converting from Standard Notation to Scientific Notation

Step 1: Place the decimal point such that there is one nonzero digit to the left of the decimal point.

Step 2: Count the number of decimal places that the decimal has "moved" from the original number. This will determine the exponent of the 10.

Step 3: If the original number is less than 1, the exponent is negative; if the original number is greater than 1, the exponent is positive.

EXAMPLE 1

Convert 5268 from standard notation to scientific notation.

$$5268 = 5268.0 = 5.268 \cdot 10^3$$
$$\underset{3\ 2\ 1}{}$$

To get one nonzero digit to the left of the decimal place, you will need to move the decimal point (which is understood to be at the far right of the number to start out) three places to the left. Therefore, the exponent is 3. This also means that the number a in this example is 5.268. Since the original number was greater than 1, the exponent in scientific notation is positive. Thus, the answer: $5.268 \cdot 10^3$.

EXAMPLE 2

Convert 0.00689 to scientific notation.

$$0.00689 = 006.89 = 6.89 \cdot 10^{-3}$$
$$\underset{1\ 2\ 3}{}$$

Converting from Scientific Notation to Standard Notation

Step 1: If the exponent is positive, move the decimal point to the right. The exponent tells you how many places to move the decimal point.

Step 2: Or, if the exponent is negative, move the decimal point to the left. Again, the exponent tells you how many places to move the decimal point.

EXAMPLE 4

Convert $7.21 \cdot 10^3$ from scientific notation to standard form.

$$7.21 \cdot 10^3 = 7210$$

Because the exponent is positive, move the decimal point to the right three places. The resulting number, 7210, is greater than 1 as expected.

Scientific Notation on Graphing Calculators

Most calculators, including the TI-83 and TI-84 models, will automatically express very small and very large numbers in scientific notation. In order to understand the display, consider the following example.

If the calculator display shows:

2.679E2

The number following the "E" is the exponent of 10. The full translation is:

$$2.679E2 = 2.679 \cdot 10^2$$

EXAMPLE 5

Write the calculator display 5.697E−4 in scientific and standard notation.

$$5.697 \cdot 10^{-4} = 0.0005697$$

Because the exponent is negative, move the decimal point to the left four places. You must put zeroes in as "place holders" as needed. The resulting answer of 0.0005697 is less than 1 as expected.

3.5 PRACTICE PROBLEMS

1. Fill in the following table by making all the necessary conversions.

Standard Notation	Scientific Notation
560,000	
	$2.8946 \cdot 10^2$
	$6.5 \cdot 10^{-5}$
123.89	
	$2.7 \cdot 10^{-4}$
0.1589	
	$5.97 \cdot 10^8$
0.00057	
0.00000079	

2. Put the following calculator displays into scientific notation and then standard notation.

Calculator Display	Scientific Notation	Standard Notation
3.265E1		
2.79E−5		
4.587E6		
5.222E−8		

CHAPTER 4
ORDER OF OPERATIONS, EXPRESSIONS, AND PROBABILITY

4.1 Adding Decimals in Discrete Probability Distributions

As you begin to work with binomial and normal probabilities, you must add, subtract, and compare decimal numbers which may or may not have the same number of decimal places.

Adding Decimals

First let's consider addition of decimals. To add decimal numbers, you need to line up the decimal point and then add each column beginning from the right-most place, carrying over when necessary.

EXAMPLE 1

What is the sum of 0.15 and 0.1678?

Step 1: Place the numbers vertically with the decimal places aligned:

$$0.15$$
$$\underline{0.1678}$$

Step 2: For some students this may seem easier if zeros are placed after the 5 in the first number in order to make the numbers the same length.

$$0.1500$$
$$\underline{0.1678}$$

Step 3: Add the columns as customary in addition:

$$\begin{array}{r} 0.1500 \\ +\ \underline{0.1678} \\ 0.3178 \end{array}$$

4.1 PRACTICE PROBLEMS

1. Add the following decimals.

   a. $0.36 + 0.47$

   b. $0.117 + 0.2$

   c. $0.5 + 0.499$

Use the probability distribution shown below in answering questions 2 and 3.

x	P(x)
0	0.1642
1	0.3462
2	0.3041
3	0.1424
4	0.0375
5	0.0053
6	0.0003

2. What is the sum of the probabilities?

3. Find the sum of probabilities, *P(1) + P(2) + P(3)*.

4. For normally distributed heights among adult women, the probability that a randomly selected woman is under 5 feet tall is 0.1007. The probability that a randomly selected woman is over 5′10″ (70 inches) tall is 0.0016. Add the values together to determine the probability of randomly selecting a woman either under 5 feet or over 5′10″.

   0.1007 + 0.0016 =

4.2 Evaluating the Mean of a Discrete Probability Distribution

Formula for Mean

When calculating the mean for a probability distribution the formula is:

$$\mu = \sum \left(x \cdot P(x) \right)$$

Because multiplication comes before addition in the order of operations, there is an implied parenthesis after the summation (Σ) sign. This indicates that we should multiply each value of the random variable, x, by its probability, $P(x)$ and then add all of these products together.

EXAMPLE 1

In one of Mendel's genetic experiments with peas, the probability of a pea plant having green pods was found to be 75%. The probability distribution shown below is for a group of six plants.

Green Pods, x	Probability $P(x)$
0	0.000
1	0.004
2	0.033
3	0.132
4	0.297
5	0.356
6	0.178

Step 1: Multiply each value of x times its probability. For clarity this step is shown as an additional column on the distribution table.

Green Pods, x	Probability $P(x)$	$x \cdot P(x)$
0	0.000	$0 \cdot 0.000 = 0.000$
1	0.004	$1 \cdot 0.004 = 0.004$
2	0.033	$2 \cdot 0.033 = 0.066$
3	0.132	$3 \cdot 0.132 = 0.396$
4	0.297	$4 \cdot 0.297 = 1.188$
5	0.356	$5 \cdot 0.356 = 1.780$
6	0.178	$6 \cdot 0.178 = 1.068$

Step 2: Now perform the summation, $\Sigma(x \cdot P(x))$, of the products found in step 1.

$\mu = \Sigma(x \cdot P(x)) = 0.000 + 0.004 + 0.066 + 0.396 + 1.188 + 1.78 + 1.068 = 4.502$

The mean is 4.502 plants out of six will have green pods.

4.2 PRACTICE PROBLEMS

1. Verify that the distribution shown below is a probability distribution by demonstrating that the probabilities sum to 1. Show your work.

x	P(x)
0	0.011
1	0.03
2	0.138
3	0.27
4	0.317
5	0.187
6	0.047

2. Find the mean of the distribution above using the formula $\mu = \Sigma(x \cdot P(x))$. Show all your work.

3. Verify that the distribution shown below is a probability distribution by demonstrating that the probabilities sum to 1. Show your work.

x	P(x)
0	0.145
1	0.342
2	0.322
3	0.152
4	0.036
5	0.003

4. Find the mean of the distribution above using the formula $\mu = \Sigma(x \cdot P(x))\ \mu = \Sigma(x \cdot P(x))$. Show all your work.

4.3 Subtracting Decimals in Discrete Probability Distributions

Subtracting Decimals

As seen in the addition of probabilities in discrete distributions, it is also critical in subtraction of probabilities to align the decimal point and place values. You should also add zeroes as needed to make the decimal numbers the same length.

EXAMPLE 1

What is the difference between 0.694 and .05?

Step 1 and 2: Make the numbers the same length and align the decimal points.

$$
\begin{array}{r}
0.694 \\
\underline{0.050} \\
\end{array} \longleftarrow
\begin{array}{l}
\text{A zero has been added in the} \\
\text{thousandths place to make them} \\
\text{the same length.}
\end{array}
$$

Step 3: Now subtract as customary.

$$
\begin{array}{r}
0.694 \\
-0.050 \\
\hline
0.644 \\
\end{array}
$$

Complements

As we discovered earlier, the probability of an event and its complement must sum up to 1. Therefore, when considering problems concerning complementary events we may need to subtract a known probability from 1. This is easy to do if we add a decimal point to the right of 1 and then insert enough zeros to make this number the same length as the probability that is to be subtracted.

EXAMPLE 2

Statistics show that 68% of high school students reported cheating on at least one exam during their years in school. What is the probability that a randomly selected student has NOT cheated on an exam during their years in school?

We need to recognize that we were given the information that P (*cheated*) = 0.68. To find the probability that the student has never cheated, we need to subtract the probability that they have cheated from 1. P (*cheated*) = 0.68.

$$P(cheated) + P(has\ never\ cheated) = 1$$

Step 1: Note that the given probability has two decimal places. So we need to add a decimal point and 2 zeros to the right of 1, making it 1.00.

Step 2: Now line the decimal places up and subtract each column, borrowing and carrying between columns as necessary.

$$
\begin{array}{r}
1.00 \\
-0.68 \\
\hline
0.32
\end{array}
$$

Since the sum of the probabilities of all the possible outcomes must be 1, we sometimes find that the easiest way to a solution is through the use of the complement.

EXAMPLE 3

We know that the probability of a woman being 65 inches or less in height equals 0.6772. What is the probability that a randomly selected woman will be over 65 inches in height?

The event of being less than or equal to 65 inches is the complement of being over 65 inches tall. Therefore, we can find the probability by subtracting the probability from 1.

Step 1: Note that the given probability has four decimal places. So we need to add a decimal point and 4 zeros to the right of 1, making it 1.0000.

Step 2: Now line the decimal places up and subtract.

$$
\begin{array}{r}
1.0000 \\
-0.6772 \\
\hline
0.3228
\end{array}
$$

The probability that a randomly selected woman is over 65 inches in height is 0.3228.

4.3 PRACTICE PROBLEMS

1. Subtract the following decimals.

   a. $1 - 0.89$

   b. $0.46 - 0.12$

x	P(x)
0	0.1642
1	0.3462
2	0.3041
3	0.1424
4	0.0375
5	0.0053
6	0.0003

2. Using the table above, subtract to find the probability the outcome is *not* 3.

$$1 - P(3) =$$

3. The ages of children in a dataset is normally distributed. The probability of randomly selecting a child under age 12 is 0.6544 while the probability of randomly selecting a child under 15 is 0.886. The probability of a value between 12 and 15 would be the difference between these two numbers.

$$0.886 - 0.6544 =$$

4.4 Interpreting and Using Inequality Notation for Discrete Random Variables

Inequality Symbols

An inequality statement is used to indicate a range of values in an expression. The following table lists some of the inequality and equality symbols and their meaning(s).

Symbol	Meaning	Equivalent Phrases
$<$	Less than	"fewer," "under," "below," or "smaller"
$>$	Greater than	"more," "over," "above," or "bigger"
$\leq$	Less than or equal to	"At most" or "no more than"
$\geq$	Greater than or equal to	"At least" or "no fewer than"
$=$	Equal to	"Is," "no different," "same as"

We encounter inequality conditions often when we work with probability. Consider the following survey in which it was found that 60% of adults believe in the devil. When we randomly select five adults and ask if they believe in the devil, the probability distribution shown below summarizes the results:

x	P(x)
0	0.010
1	0.077
2	0.230
3	0.346
4	0.259
5	0.078

Let's consider a few examples which use this probability distribution and the logic associated with inequality phrases. We will draw a number line to represent the probability statements—this number line will only include numbers from 0 to 5, since those are the only possible values from this survey. Further, because we are considering people surveyed, only whole numbers are appropriate.

EXAMPLE 1

Find the probability that more than 4 of the five adults surveyed say they believe in the devil.

The phrase "more than 4" is important to consider. Ask yourself, does this include 4? The answer should be "no" because it specifically says <u>more</u> than 4. Therefore, if we represented this phrase on a number line it would not include 4 and would look like this:

Note that the number 4 is excluded. Only numbers greater than 4 are plotted. The symbol for greater than is > so the probability notation should be:

$$P(more\ than\ 4) = P(x > 4)$$

EXAMPLE 2

Find the probability that 1 or fewer in the group of five say they believe in the devil.

The phrase "1 or fewer" is important to consider. Ask yourself, does this include 1? The answer is "yes" and it also includes 0 because this is less than 1. Therefore, if we represented this phrase on a number line it would look like this:

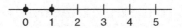

The symbolic notation needs to include 1 or fewer which is the symbol ≤. This probability then would be:

$$P(1\ or\ less) = P(x \leq 1) = P(0) + P(1)$$

EXAMPLE 3

Find the probability that no more than 4 of the five say they believe in the devil.

The phrase "no more than 4" is important to consider. Ask yourself, does this include 4? The answer is "yes" and it also includes all whole numbers less than 4. Therefore, if we represented this phrase on a number line it would look like this:

This is the sum of the probabilities from 0 to 4.

$$P(no\ more\ than\ 4) = P(x \leq 4) = P(0) + P(1) + P(2) + P(3) + P(4)$$

4.4 PRACTICE PROBLEMS

A college admissions committee is reviewing the ACT scores of its applicants to determine the appropriate cut-off points for entering freshmen. Write the following statements as inequalities by filling in the correct symbol.

1. ACT scores of 19 and lower

$$x \boxed{} 19$$

2. Scores between 26 and 30

$$26 \boxed{} x \boxed{} 30$$

3. ACT scores over 12 and at most 18

$$12 \boxed{} x \boxed{} 18$$

4. Scores above 21 on the ACT

$$21 \boxed{} x$$

The probability of inheriting a certain genetic mutation has been shown to be 0.25. Nancy and Bill have four children. The probability distribution for the inheritance of this disorder is given by the following table:

x	P(X)
0	0.316
1	0.422
2	0.211
3	0.047
4	0.004

Write the appropriate inequality using probability notation for each of the statements below. You do not need to determine the probability value.

5. The probability that no more than one of the children inherit the genetic disorder:

6. The probability that at least one of the children inherit the genetic disorder:

7. The probability that three or fewer of the children inherit the genetic disorder:

8. The probability that all of the children inherit the genetic disorder:

UNIT III
PRINCIPLES OF CONTINUOUS PROBABILITY DISTRIBUTIONS

CHAPTER 5
INEQUALITIES, AREA, AND CONTINUOUS DISTRIBUTIONS

5.1 Interpreting and Using Inequality Notation for Continuous Random Variables

Inequality Notation and Continuous Probability

Unlike the discrete probability distributions, continuous random variables can take any real number value within a given range. Therefore, their representation on a number line is different than that seen for a discrete random variable. Closed circles indicate an included value, whereas open circles strictly do not include the number.

EXAMPLE 1

What is the representation of a woman's height between 58 and 62 inches tall?

To express this statement as an inequality, we need to consider the meaning of the word "between." Specifically, does the phrase "between 58 and 62" include these numbers? Correctly interpreted, the phrase would not include 58 or 62, only the numbers that are literally between them. On a number line this phrase "between 58 and 62" looks like this:

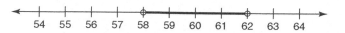

Written as an inequality this would be $58 < x < 62$ where the value of x can only be numbers between, but not including, 58 and 62.

If you want to include the endpoints, 58 and 62, in the inequality, then it must be stated differently. If you ask "what is the representation of a woman's height between 58 and 62 inches tall, inclusive" then you do include the endpoints. The key word here is "inclusive" which means to include the endpoints in the interval. Now the number line would look like this:

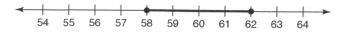

The inequality form would be $58 \leq x \leq 62$. So x can take any value between 58 and 62, including the endpoints of 58 and 62.

EXAMPLE 2

How would a score below 70 on an exam be represented on a number line and written symbolically?

This statement includes any test score below 70 but not 70 itself. Therefore, the number line representation would be:

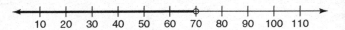

The inequality form would be $x \geq 70$ where x is the random variable for test scores.

EXAMPLE 3

How would a score of at least 70 on an exam be represented on a number line and written symbolically?

This is different from the previous example because a score of 70 or anything higher would be included in this expression. So the number line representation would be:

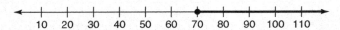

The inequality form would be $x \geq 70$ because the inequality symbol for "at least" is $\geq$.

5.1 PRACTICE PROBLEMS

For the following problems, interpret the statements as both number line graphs and symbolic statements.

1. Grades on a certain test lying between 80 and 92:

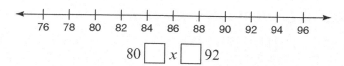

2. An egg weighing 4 ounces or more:

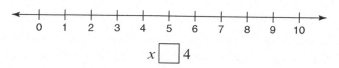

3. An egg weighing more than 4 ounces:

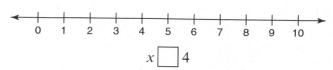

4. An IQ of at most 130:

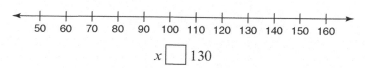

5. An IQ between 100 and 130, inclusive:

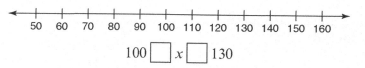

5.2 Calculating Probability using Area of Rectangles

Area of Rectangles in Uniform Distributions

Finding the area of a rectangle is necessary when you need to determine the probability of a uniform distribution. To find the area of a rectangle, multiply the length of the rectangle times its width.

$$A = L \cdot W$$

EXAMPLE 1

Wait times in line at a bank are uniformly distributed between 0 and 5 minutes. If the wait time for any customer exceeds 5 minutes, another teller is added. We wish to find the probability of waiting less than 2 minutes.

This probability rectangle is shown in the picture.

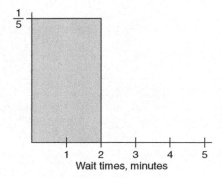

The area of the rectangle shown is equivalent to the probability of waiting for less than 2 minutes in line. Therefore, using the formula for area:

$$A = L \cdot W = 2 \cdot \frac{1}{5} = \frac{2}{5} = 0.40$$

That is, you have a 40% chance of waiting less than 2 minutes.

EXAMPLE 2

Continuing from above, find the probability that your wait time is greater than 2 minutes.

The probability rectangle is shown below:

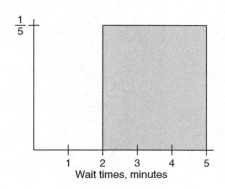

Again, using the formula for rectangular area:

$$A = L \cdot W = 3 \cdot \frac{1}{5} = \frac{3}{5} = 0.60$$

That is, you have a 60% chance of waiting more than 2 minutes. (Did you notice? The events in Examples 1 and 2 are complements. This is why their probability will sum to 1!)

5.2 PRACTICE PROBLEMS

1. The wait time for a certain bus is uniformly distributed between zero and ten minutes as shown below:

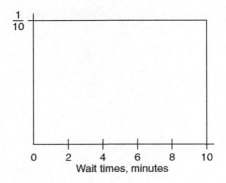

Find the probability of waiting 6 minutes or more by shading in the rectangle representing this statement on the graph above and finding its area.

2. The wait time for a certain bus is uniformly distributed between zero and ten minutes as shown below:

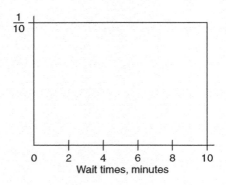

Find the probability of waiting less than 4 minutes by shading in the rectangle representing this statement on the graph above and finding its area.

3. Consider the pin-the-tail-on-the-donkey target shown below. If you know that the donkey figure has an area of 2.5 ft^2 and that any pin stuck in the donkey scores one point in the game, what is the probability that you will score a point?

$$Probability = \frac{area\ of\ the\ donkey}{total\ area\ of\ picture}$$

2 ft

3 ft

5.3 Calculating Probability Using Other Geometric Shapes

Triangles

The formula for the area of a triangle is

$$\text{Area} = \frac{1}{2} \cdot b \cdot h$$

where b is the base and the height, h, must be perpendicular to the base. We often draw the picture:

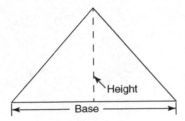

EXAMPLE 1
Consider a triangle whose base is 7 in. and whose height is 2 in.

$$\begin{aligned}
\textit{Area} &= \frac{1}{2} \cdot b \cdot h \\
&= \frac{1}{2} \cdot 7 \text{ in} \cdot 2 \text{ in} \\
&= 7 \text{ in}^2
\end{aligned}$$

Notice that the units are squared.

EXAMPLE 2
Consider the triangle shown below.

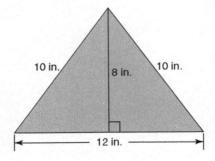

The 12 inch side is perpendicular (at right angles) to the 8 inch line. The 8 inch line is the height of the triangle even though it is not an actual side of the triangle. Substituting into the formula:

$$Area = \frac{1}{2} \cdot b \cdot h$$

$$= \frac{1}{2} \cdot 12 \text{ in} \cdot 8 \text{ in}$$

$$= 48 \text{ in}^2$$

Circles

For circles the area is defined by this formula:

$$\text{Area} = \pi \cdot r^2$$

where the irrational number, π, is often estimated by the number 3.14 and r is the radius of the circle. Remember that the radius of a circle is also one half of its diameter.

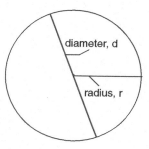

EXAMPLE 3

Find the area of the circle shown below:

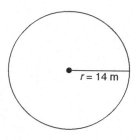

$$Area = \pi \cdot r^2$$

$$= \pi \cdot (14 \text{ m})^2$$

$$= 196 \, \pi \approx 615.44 \text{ m}^2$$

Again, notice that the units are square meters here. Also note that 615.44 meters squared is an approximation because we have used an estimate for the value of pi, π.

EXAMPLE 4

Find the area of the circle shown below. Give an exact answer in terms of π.

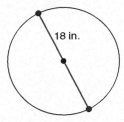

The radius is half of the diameter. Therefore, the radius of this circle is 9 inches.

$$Area = \pi \cdot r^2 = \pi \cdot (9 \text{ in})^2 = 81\pi \text{ in}^2$$

5.3 PRACTICE PROBLEMS

1. Find the area of the figure shown. Show all of your work.

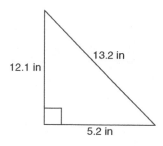

2. Find the area of the figure shown. Show all of your work.

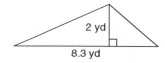

3. Find the area of the figure shown. Show all of your work.

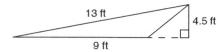

4. Find the area of the figure shown. Show all of your work. Use the estimate $\pi \cong 3.14$.

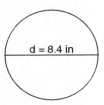

5. Find the area of the figure shown. Show all of your work. Use the estimate $\pi \cong 3.14$.

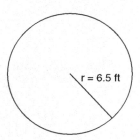

6. Consider the dart board shown below. What is the probability that I will hit the inner circle of the target? Show all of your work.

$$Probability = \frac{area\ of\ inner\ circle}{area\ of\ outer\ circle}$$

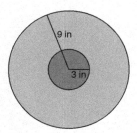

7. Consider the dart board shown below. What is the probability that I will hit the triangle in the rectangular target? Show all of your work.

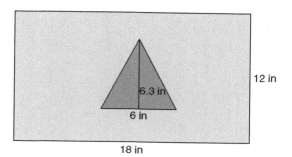

CHAPTER 6

LINEAR EQUATIONS, INEQUALITIES, AND FORMULAS

6.1 Solving Two-Step Linear Equations

Solving linear equations is the task of isolating the variable of interest, often x, and finding the value of this variable. To isolate the variable on one side of the equal sign, you must keep in mind the following:

1. Always start on the side of the equation with the variable.

2. What operations you do to one side of the equal sign, you must do to the other side as well.

EXAMPLE 1 Solve for x.

$$5x - 1 = 19$$

Step 1: Addition and subtraction are operations that "undo" each other. To remove the subtracted 1 on the left side, we must add 1 to BOTH sides.

$$5x - 1 + 1 = 19 + 1$$
$$5x = 20$$

Step 2: Multiplication and division are also operations that "undo" each other. To remove the 5 which is multiplying x, we must divide BOTH sides by 5.

$$\frac{5x}{5} = \frac{20}{5}$$
$$x = 4$$

EXAMPLE 2 Solve for x.

$$\frac{x}{2} + 5 = 35$$

Step 1: Eliminate the 5 by subtracting it from both sides.

$$\frac{x}{2} + 5 - 5 = 35 - 5$$
$$\frac{x}{2} = 30$$

95

Step 2: Eliminate the 2 (which is currently dividing the x variable) by multiplying by 2 on both sides.

$$\frac{x}{2} \cdot 2 = 30 \cdot 2$$

$$x = 60$$

6.1 PRACTICE PROBLEMS

Solve problems 1–10 using the two-step process shown above. Show all of your work.

1. $3x + 7 = 19$

2. $\dfrac{x}{5} - 13 = 4$

3. $2x - 8 = 15$

4. $\dfrac{1}{2}x - 3 = 5$

5. $2z - \dfrac{2}{3} = \dfrac{7}{6}$

6. $2x - 12 = 5$

7. $67 = 11y + 1$

8. $7 = 4 - 2x$

9. $3y - 9 = 14$

10. $\dfrac{2}{3}x + 4 = 13$

6.2 Solving Multi-Step Linear Equations

Only some linear equations in one variable can be solved with two steps. Linear equations which have the variable on both sides of the equation, for example, require additional steps. The additional steps often involve one or more of the following:

1. Use addition or subtraction of terms to get the variable on the same side.

2. Combine the variable terms into a single term.

3. Perform distribution when the variable is in a parenthesis.

Consider the following examples.

EXAMPLE 1   Solve for x.

$$4x - 3 = 22 - x$$

Step 1: Since there is a variable term on both sides of the equation, we need to use addition to move one of them to the opposite side. In this example we will move the variable term from the right side to the left side by addition of an "x" to BOTH sides.

$$4x - 3 + x = 22 - x + x$$

$$4x - 3 + x = 22$$

Step 2: Combine $4x$ and x, which are called "like terms," on the left to further simplify.

$$5x - 3 = 22$$

Step 3: Eliminate the subtracted term on the left by adding 3 to both sides.

$$5x - 3 + 3 = 22 + 3$$

$$5x = 25$$

Step 4: Eliminate the 5, which is multiplying the x value, by dividing both sides by 5.

$$\frac{5x}{5} = \frac{25}{5}$$

$$x = 5$$

Distribution

When there is a number outside a parenthesis, it indicates that each of the terms inside the parentheses must be multiplied by that number. The process of rewriting an expression without the parentheses by multiplication is called distribution.

EXAMPLE 2   Solve for x.

$$6(x - 7) = 4$$

Step 1: In this problem we need to first distribute the 6 by multiplying both the x and the 7 inside the parentheses by 6.

$$6(x - 7) = 4$$

$$6 \cdot x - 6 \cdot 7 = 4$$

$$6x - 42 = 4$$

Step 2: Eliminate the subtracted term on the left by adding 42 to both sides.

$$6x - 42 + 42 = 4 + 42$$

$$6x = 46$$

Step 3: Eliminate the 6, which is multiplying the x value, by dividing both sides by 6.

$$\frac{6x}{6} = \frac{46}{6}$$

$$x = \frac{46}{6}$$

This can also be written $x = 7.67$, rounded to the nearest hundredths place.

EXAMPLE 3 Solve for x.

$$0.5(x + 4) = 2(3x - 1)$$

Step 1: First distribute to simplify both sides.

$$0.5(x + 4) = 2(3x - 1)$$

$$0.5 \cdot x + 0.5 \cdot 4 = 2 \cdot 3x - 2 \cdot 1$$

$$0.5x + 2 = 6x - 2$$

Step 2: Move the variable term from the right side to the left side.

$$0.5x + 2 - 6x = 6x - 2 - 6x$$

$$0.5x + 2 - 6x = -2$$

Step 3: Simplify the left hand side by combining like terms.

$$0.5x + 2 - 6x = -2$$

$$-5.5x + 2 = -2$$

Step 4: Subtract 2 from both sides to eliminate it from the left side.

$$-5.5x + 2 - 2 = -2 - 2$$

$$-5.5x = -4$$

Step 5: Isolate the variable by dividing both sides by -5.5. Notice dividing a negative number by a negative number results in a positive quotient.

$$\frac{-5.5x}{-5.5} = \frac{-4}{-5.5}$$

$$x = \frac{4}{5.5}$$

This can also be written $x = 0.727$, rounded to the nearest thousandths place.

6.2 PRACTICE PROBLEMS

Solve for the variable. Show all of your work.

1. $7x - 8 - 9x = -14$

2. $2x + 5x = 21$

3. $3z + 6 - 2z = 14 - 7$

4. $-9r + 6 = 9r + 24$

5. $18 - 4x - 4 = 7x + 5x$

6. $-3(x + 2) = 12 + 3$

7. $\dfrac{3(r-5)}{5} = 3r$

8. $8x - 1 = 21 - 3x$

9. $5x + 4 - 7 = 4x - 2$

10. $0.2x + 0.7 = 1.4 - 0.1x$

6.3 Solving Proportions that Simplify to Linear Expressions

A proportion is an equation which states that two ratios (fractions) are equal to each other. If one term of a proportion is not known (usually designated by the variable x), the concept of cross multiplication can be used to find the value of the unknown term.

Here is an example of a proportion:

$$\frac{x}{3} = \frac{2}{9}$$

Cross Multiplication

To cross multiply, you take the left side fraction's numerator and multiply it by the right side fraction's denominator. Then take the left side fraction's denominator and multiply it by the right side fraction's numerator. These two terms are set equal to one another. Now you have eliminated the fractions and can solve as you would for other linear equations.

Consider the examples below.

EXAMPLE 1   Solve the proportion for x:

$$\frac{x}{3} = \frac{2}{9}$$

Step 1: Cross multiply.

$$\frac{x}{3} \bowtie \frac{2}{9}$$

$$9 \cdot x = 3 \cdot 2$$
$$9x = 6$$

Step 2: Divide both sides by 9.

$$\frac{9x}{9} = \frac{6}{9}$$

$$x = \frac{6}{9}$$

This fraction can be reduced to $x = \dfrac{2}{3}$.

EXAMPLE 2   Solve the proportion for x:

$$\frac{4}{12} = \frac{9}{x}$$

Step 1: Cross multiply:

$$4 \cdot x = 12 \cdot 9$$
$$4x = 108$$

Step 2: Divide both sides by 4.

$$\frac{4x}{4} = \frac{108}{4}$$
$$x = 27$$

EXAMPLE 3   Solve the proportion for x:

$$\frac{3}{8} = \frac{6}{x+4}$$

Step 1: Cross multiply. Be sure to add parentheses around $x+4$ to preserve the order of operations.

$$3 \cdot (x+4) = 8 \cdot 6$$
$$3(x+4) = 48$$

Step 2: Distribute:

$$3 \cdot x + 3 \cdot 4 = 48$$
$$3x + 12 = 48$$

Step 3: Subtract 12 from both sides.

$$3x + 12 - 12 = 48 - 12$$
$$3x = 36$$

Step 4: Divide both sides by 3.

$$\frac{3x}{3} = \frac{36}{3}$$
$$x = 12$$

6.3 PRACTICE PROBLEMS

Solve the following proportions.

1. $\dfrac{3}{4} = \dfrac{x}{8}$

2. $\dfrac{5}{x} = \dfrac{4}{x+2}$

3. $\dfrac{2(x+1)}{4} = \dfrac{5}{2}$

4. $\dfrac{4x}{7} = \dfrac{2}{3}$

5. $\dfrac{2x+1}{2} = \dfrac{3}{5}$

6. $\dfrac{x+4}{7} = \dfrac{x-1}{5}$

7. $\dfrac{2}{5} = \dfrac{x}{10-x}$

8. $\dfrac{9}{x+3} = \dfrac{11}{5-x}$

6.4 Solving Linear Inequalities and Graphing the Solution

Linear inequalities in one variable can be solved in much the same fashion as linear equations. The general steps include:

1. Use addition or subtraction of terms to get the variables on the same side.

2. Combine the variable terms into a single term.

3. Perform distribution when the variable is in a parenthesis.

4. Use multiplication and division to simplify.

However, inequalities have one rule that is particular to their solution:

5. **If you multiply or divide by a negative number, the direction of the inequality sign must be reversed.**

The logic of this rule can be shown using a simple example.

We know that $4 < 5$ which means the number 4 is smaller than the number 5. Look what happens when we multiply both sides of this inequality by -1.

$$4 \cdot (-1) < 5 \cdot (-1)$$

Results in:

$$-4 < -5$$

But, we know by looking at the number line that -5 is actually farther to the left, and therefore less than -4, not the other way around. We must flip the direction of the inequality to make this correct.

$$-4 > -5$$

Consider some examples in solving inequalities.

EXAMPLE 1 Solve the inequality for x and graph on a number line.

$$4x + 2 > 10$$

Step 1: Isolate the x term by subtracting 2 from both sides.

$$4x + 2 - 2 > 10 - 2$$
$$4x > 8$$

Step 2: Divide both sides by 4. There is no need to flip the direction of the inequality because 4 is a positive number.

$$\frac{4x}{4} > \frac{8}{4}$$

$$x > 2$$

Step 3: We need to show on the number line that x can take any value greater than, but not equal to, 2.

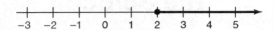

EXAMPLE 2   Solve the inequality for x and graph on a number line.

$$-2x+5 \leq x+14$$

Step 1: Subtract x from both sides in order to get all the x terms on the left.

$$-2x+5-x \leq x-x+14$$
$$-2x-x+5 \leq 14$$

Step 2: Combine like terms.

$$-3x+5 \leq 14$$

Step 3: Subtract 5 from both sides.

$$-3x+5-5 \leq 14-5$$
$$-3x \leq 9$$

Step 4: Divide both sides by -3, remembering to flip the direction of the inequality.

$$\frac{-3x}{-3} \leq \frac{9}{-3}$$
$$x \geq -3$$

Step 5: On the number line graph the values of x that are greater than or equal to -3.

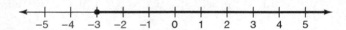

EXAMPLE 3   Solve the inequality for x and graph on a number line.

$$-2(x-5) > 2(x+3)$$

Step 1: Distribute across the parentheses on both sides.

$$-2x+10 > 2x+6$$

Step 2: Subtract $2x$ from both sides.

$$-2x-2x+10 > 2x-2x+6$$

Step 3: Combine like terms.

$$-4x+10 > 6$$

Step 4: Subtract 10 from both sides.

$$-4x+10-10 > 6-10$$
$$-4x > -4$$

Step 5: Divide both sides by -4, remembering to flip the direction of the inequality.

$$\frac{-4x}{-4} > \frac{-4}{-4}$$

$$x < 1$$

Step 6: On the number line graph the values of x that are less than 1.

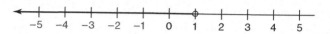

6.4 PRACTICE PROBLEMS

Solve the linear inequality and graph the solution on the provided number line.

1. $2x - 3 > 8$

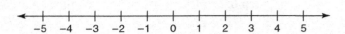

2. $-x + 4 \leq 7$

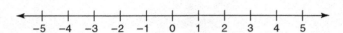

3. $2(y - 1) \geq 3$

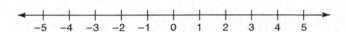

4. $x + 2 \leq 2x + 5$

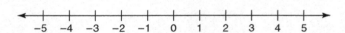

5. $3 < 2x + 3 \leq 5$

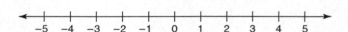

6. $-14 < -2x - 6 < -4$

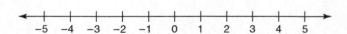

6.5 Solving Literal Equations for a Given Variable

Literal Equations

A literal equation is one that involves several variables or symbols. In statistics, many of the formulas are literal equations, such as the formula for the mean test statistic:

$$t = \frac{\overline{x} - \mu}{\dfrac{s}{\sqrt{n}}}$$

Note that all of the values in this equation are symbolic. Refer to your statistics text to determine the meaning of each of them.

These types of equations can be solved for any of the variables involved. This might be useful if we need to know the value of a variable different from the one already isolated in the formula. To do this, we treat the variables as if they are "just numbers" and use the same steps we used to solve linear equations. Consider the following examples.

EXAMPLE 1   In the formula for the perimeter of a rectangle, solve for the length, l.

$$P = 2w + 2l.$$

Step 1: We wish to solve for the length, l, so we need to isolate it to one side. Begin by subtracting $2w$ from both sides.

$$P - 2w = 2w + 2l - 2w$$

$$P - 2w = 2l$$

Step 2: Now divide both sides by 2.

$$\frac{P - 2w}{2} = \frac{2l}{2}$$

$$l = \frac{P - 2w}{2}$$

EXAMPLE 2   Solve the margin of error formula for the sample size, n.

$$E = z \cdot \frac{\sigma}{\sqrt{n}}$$

Step 1: We wish to isolate n, so first we need to multiply both sides by $\sqrt{n}$, to remove the variable from the denominator.

$$\sqrt{n} \cdot E = z \cdot \frac{\sigma}{\sqrt{n}} \cdot \sqrt{n}$$

$$\sqrt{n} \cdot E = z \cdot \sigma$$

Step 2: Now divide both sides by E, again to isolate the variable n.

$$\frac{\sqrt{n} \cdot E}{E} = \frac{z\sigma}{E}$$

$$\sqrt{n} = \frac{z\sigma}{E}$$

Step 3: To eliminate the square root, similar to addition/subtraction or multiplication/division, we will need to do the "opposite" operation, which is raising both sides to the 2nd power.

$$\left(\sqrt{n}\right)^2 = \left(\frac{z\sigma}{E}\right)^2$$

$$n = \left(\frac{z\sigma}{E}\right)^2$$

This can also be written:

$$n = \frac{z^2 \cdot \sigma^2}{E^2}$$

6.5 PRACTICE PROBLEMS

1. The formula for the volume of a rectangular solid is given by $V = lwh$.
   Solve this formula for the width, w.

2. The formula for the volume of a cylinder is given by $V = \pi r^2 h$.
   Solve this formula for the radius, r.

3. The formula for the area of a triangle is given by $A = \dfrac{1}{2}bh$.
   Solve this formula for the base, b.

4. The formula for the z-score for the value x is given $z = \dfrac{x - \mu}{\sigma}$.
   Solve this formula for the value of x.

5. What is the relationship between the two literal equations in problem 4?

UNIT IV
PRINCIPLES OF ESTIMATION AND HYPOTHESIS TESTING

CHAPTER 7

OPERATIONS, EXPRESSIONS, AND CONFIDENCE INTERVALS

7.1 Order of Operations in the Margin of Error

The complexity of the margin of error and other related formulas require a more sophisticated application of the order of operations. This includes implied parentheses and exponent manipulation.

Formula for the margin of error for a proportion:

$$E = z_{\alpha/2} \cdot \sqrt{\frac{\hat{p} \cdot \hat{q}}{n}} = z_{\alpha/2} \left(\frac{\hat{p} \cdot \hat{q}}{n} \right)^{\frac{1}{2}}$$

Formula for the margin of error for a mean, population standard deviation unknown:

$$E = t_{\alpha/2} \cdot \frac{s}{\sqrt{n}}$$

Consider the following examples.

EXAMPLE 1

Find the margin of error, rounded to the nearest thousandth, when the sample proportion $\hat{p}$ is 0.60, the critical value $z_{\alpha/2}$ is 1.645, and the sample size n is 36.

> **Step 1:** Because $\hat{p}$ and $\hat{q}$ are complements, their sum must be 1. So:

$$\hat{q} = 1 - \hat{p} = 1 - 0.6 = 0.4$$

> **Step 2:** Recall that the square root is equivalent to raising a quantity to the $\frac{1}{2}$ power.

$$E = z_{\alpha/2} \cdot \sqrt{\frac{\hat{p} \cdot \hat{q}}{n}} = z_{\alpha/2} \left(\frac{\hat{p} \cdot \hat{q}}{n} \right)^{\frac{1}{2}}$$

> Now substitute into the equation for margin of error.

$$E = 1.645 \cdot \left(\frac{0.6 \cdot 0.4}{36} \right)^{\frac{1}{2}}$$

> **Step 3:** Do the multiplication and division inside the parentheses.

$$E = 1.645 \cdot \left(\frac{0.24}{36} \right)^{\frac{1}{2}} = 1.645 \cdot (0.006667)^{\frac{1}{2}}$$

117

Step 4: Next, carry out the exponent operation.

$$E = 1.645 \cdot 0.08165$$

Step 5: Perform the final multiplication and round to three decimal places.

$$E = 0.134$$

EXAMPLE 2

Evaluate the margin of error for a confidence interval for the mean, population standard deviation unknown, where the sample standard deviation s is 2.68, the t critical value is 2.086, and the size of the sample n is 21.

Step 1: Substitute the given values into the margin of error formula.

$$E = t_{\alpha/2} \cdot \frac{s}{\sqrt{n}} = 2.086 \cdot \frac{2.68}{\sqrt{21}}$$

Step 2: Perform the exponent operation by taking the square root of the sample size.

$$E = 2.086 \cdot \frac{2.68}{4.583}$$

Step 3: Moving from left to right, carry out the multiplication and division operations.

$$E = 1.22$$

7.1 PRACTICE PROBLEMS

Use the order of operations to calculate the margin of error for each problem. Where requested, identify the variables. Show all of your work.

1. On a recent promotion exam, 34 candidates passed the exam out of a total of 77 candidates. At the 95% confidence level ($z_{\alpha/2} = 1.96$), calculate the margin of error for the proportion who passed. Show all work.

$$\hat{p} = \frac{x}{n} =$$

$$\hat{q} =$$

$$n =$$

$$E = z_{\alpha/2} \sqrt{\frac{\hat{p} \cdot \hat{q}}{n}} =$$

2. One statistical study of 325 youth showed that 10% of them smoke. What is the margin of error for percentage of smokers, if we want to be 95% confident ($z_{\alpha/2} = 1.96$)?

3. Calculate the margin of error for a mean, with population standard deviation unknown, when the sample standard deviation is 3.59, the sample size is 51 and the t critical value is 2.009. Show all work.

$$E = t_{\alpha/2} \cdot \left(\frac{s}{\sqrt{n}} \right) =$$

4. Calculate the margin of error for mean per capita income in Appalachian counties. The critical value for a sample of 300 Appalachian counties is $t_{\alpha/2} = 1.97$ with a sample standard deviation of 4221 dollars.

7.2 Evaluating the Minimum Required Sample Size

One of the calculations we are asked to compute is the minimum required sample size needed to construct a confidence interval for a proportion or a mean.

There are two formulas for proportions, depending upon whether or not we have an estimate for the proportion. The formula for a mean can only be used in instances when the population standard deviation is known.

Formula for sample size for a proportion, with a known estimate:

$$n = \frac{\left(z_{\alpha/2}\right)^2 \cdot \hat{p} \cdot \hat{q}}{E^2}$$

Formula for sample size for a proportion, with no known estimate:

$$n = \frac{\left(z_{\alpha/2}\right)^2 \cdot 0.25}{E^2}$$

Formula for sample size for a mean, population standard deviation known:

$$n = \left(\frac{z_{\alpha/2} \cdot \sigma}{E}\right)^2$$

Important note: we ALWAYS round up because the idea is MINIMUM sample size.

EXAMPLE 1

Determine the sample size needed to estimate the current proportion of Republican voters among all registered voters in Iowa if we want to be 95% confident ($z_{\alpha/2} = 1.96$) with a margin of error no larger than 3%. We know that in the most recent election, 58% of the voters were registered as Republican.

We need to find the sample size of a proportion and need to use the formula:

$$n = \frac{\left(z_{\alpha/2}\right)^2 \cdot \hat{p} \cdot \hat{q}}{E^2}$$

From this description we can identify the following symbolic values:

$$\hat{p} = 0.58$$
$$\hat{q} = 0.42$$
$$E = 0.03$$
$$z_{\alpha/2} = 1.96$$

Step 1: Substitute the identified values for each of the variables into the equation.

$$n = \frac{1.96^2 \cdot 0.58 \cdot 0.42}{0.03^2}$$

Step 2: Following the order of operations, the exponent operations must be done first.

$$n = \frac{3.8416 \cdot 0.58 \cdot 0.42}{0.0009}$$

Step 3: Now we do the multiplication in the numerator, followed by the division.

$$n = 1039.793$$

However, we must sample WHOLE people so we round up to $n = 1040$.

EXAMPLE 2

Find the sample size needed to estimate the percentage of people who use Snapchat® at a 99% confidence level ($z_{\alpha/2} = 2.575$) and a margin of error of 2 percentage points.

Here we do not know an estimate of p. If we do not have an estimate of the proportion, then we use the formula:

$$n = \frac{\left(z_{\alpha/2}\right)^2 \cdot 0.25}{E^2}$$

We can identify the following:

$$z_{\alpha/2} = 2.575$$
$$E = 0.02$$

Step 1: Substitute into the equation.

$$n = \frac{2.575^2 \cdot 0.25}{0.02^2}$$

Step 2: Following the order of operations, the exponent operations must be done first.

$$n = \frac{6.630625 \cdot 0.25}{0.0004}$$

Step 3: Now we must do the multiplication and division operations.

$$n = 4144.14$$

For the final answer, round up to $n = 4145$.

EXAMPLE 3

Find the minimum sample size needed to estimate the mean ACT score for college applicants when we want to be 95% confident ($z_{\alpha/2} = 1.96$) with a margin of error of at most 3 points on the exam. Prior studies have shown that a good estimate of the population standard deviation σ is 6.5.

We need to estimate the sample size for mean using the formula:

$$n = \left(\frac{z_{\alpha/2} \cdot \sigma}{E} \right)^2$$

From the information we know the following:

$$z_{\alpha/2} = 1.96$$

$$\sigma = 6.5$$

$$E = 3$$

Note that the margin of error here is not a percentage but an actual number of points scored on the ACT exam.

Step 1: Substitute into the equation.

$$n = \left(\frac{1.96 \cdot 6.5}{3} \right)^2$$

Step 2: The operations inside the parentheses must be done first according to the order of operations.

$$n = 4.2466667^2$$

Step 3: Do the exponent operation.

$$n = 18.034$$

We must at least sample 19 students.

7.2 PRACTICE PROBLEMS

Use the order of operations to calculate the minimum sample size for each problem. Where requested, identify the variables. Show all of your work.

1. Determine the sample size needed to estimate the proportion of flights that are delayed due to bad weather. Data from last year suggests that 13% of flights were delayed due to weather conditions. Assume that we wish to be 95% confident ($z_{\alpha/2} = 1.96$) with a margin of error of no more than 3 percentage points.

   a. Circle the formula that should be used:

   $$n = \frac{\left(z_{\alpha/2}\right)^2 \cdot \hat{p} \cdot \hat{q}}{E^2}$$

   $$n = \frac{\left(z_{\alpha/2}\right)^2 \cdot 0.25}{E^2}$$

   b. Calculate the needed sample size using the formula selected in part a.

2. Determine the sample size needed to estimate the proportion of students who use Instagram. There is no prior estimate of Instagram use for the student body. Assume that we wish to be 95% confident ($z_{\alpha/2} = 1.96$) with a margin of error of no more than two percentage points.

3. Determine the sample size needed to estimate the proportion of broken cookies in bags of *Chips Ahoy*. Data from prior samples suggests that 4% of cookies are broken in unopened bags. Assume that we wish to be 90% confident ($z_{\alpha/2} = 1.645$) with a margin of error of no more than 2 percentage points.

4. Determine the sample size needed to estimate the mean SAT math score of entering freshmen at a local university. Data from previous years show that a good estimate of population standard deviation, σ is 8.2 points. Assume that we wish to be 95% confident ($z_{\alpha/2} = 1.96$) with a margin of error of no more 25 points on the exam.

7.3 Determining Independent and Dependent Variables

In statistics, when looking for a relationship between two variables of interest, it is customary to refer to one as the independent (also known as the predictor or explanatory) variable and one as the dependent (also known as a response) variable. For example, we may theorize that grade level is related to reading comprehension. In that case, grade level is the independent variable and reading comprehension is the dependent variable, because a student's reading comprehension depends on their grade level (and not the other way around).

Written Description:

In a written description of a problem, it is customary to mention the independent variable first, followed by the dependent variable.

Tabular Form:

When the data is presented in a table, the independent variable is the first column when in a vertical presentation or first row when in a horizontal presentation.

Cartesian Plane:

In the Cartesian (x-y) plane, the independent variable is plotted on the horizontal x-axis, and the dependent variable is plotted on the vertical y-axis.

Consider the examples below.

EXAMPLE 1

Identify the independent and dependent variables in determining if there is a linear relationship between altitude (in feet above sea level) and temperature (in degrees Fahrenheit).

Based upon the order of the variables in the description given:

> Independent variable: altitude (ft)

> Dependent variable: temperature (F°)

EXAMPLE 2

Identify the independent and dependent variables for the data set below:

Altitude (ft)	3000	8000	9000	15000	20000	25000	33000
Temperature (F°)	62	55	22	0	−10	−20	−25

Based on the row order of the horizontal table:

> Independent variable: altitude (ft)

> Dependent variable: temperature (F°)

EXAMPLE 3

Identify the independent and dependent variables in determining whether there is a linear relationship between the variables shown.

Size of diamond (carats)	Price ($)
0.2	800
0.6	1100
1.6	5500
0.5	1200
2.3	7000

Based on the column order of the vertical table:

Independent variable: size of diamond (carats)

Dependent variable: price ($)

7.3 PRACTICE PROBLEMS

1. At a routine visit to your child's pediatrician, she shows you a graph of age and weight for children between 2 and 6 years old. Identify the independent and dependent variables in this problem.

   Independent variable:

   Dependent variable:

2. Examine the graph shown below and identify the independent and dependent variables.

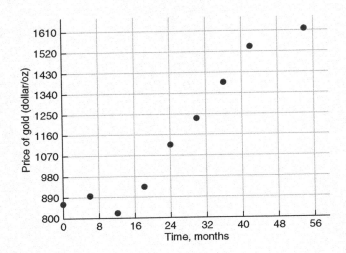

   Independent variable:

   Dependent variable:

3. Examine the data table given below and determine the independent and dependent variables.

Height (inches)	Arm Span (inches)
56	54
75	78
62	62
54	53
58	58

Independent variable:

Dependent variable:

4. Examine the data table given below and determine the independent and dependent variables.

Years of Education	8	12	20	14	16	18
Annual Salary	$24520	$25200	$52026	$34525	$48000	$48200

Independent variable:

Dependent variable:

7.4 Creating a Table of Values from a Linear Equation

To create a table of values from a known linear relationship, we need to solve the linear equation that describes the relationship for several values of the independent variable. These values of the independent variable may be given or chosen on your own in some cases.

Consider the example below.

EXAMPLE 1

The monthly cost of a certain long distance service is given by the linear equation below:

$$C = 0.07t + 3.95$$

where C is the cost in dollars and t is the number of minutes used in a month.

The independent variable in this equation is t, because the amount of the bill C, depends on how many minutes of long distance service is used. Find the cost of calling long distance for 80 minutes, 100 minutes, and 180 minutes, and complete the table.

t, total call time in minutes	C, Cost
80	$C = 0.07 \cdot 80 + 3.95 = 5.6 + 3.95 = 9.55$
100	$C = 0.07 \cdot 100 + 3.95 = 7.0 + 3.95 = 10.95$
180	$C = 0.07 \cdot 180 + 3.95 = 12.60 + 3.95 = 16.55$

So the completed table of values would look like:

t (minutes)	C (dollars)
80	$9.55
100	$10.95
180	$16.55

EXAMPLE 2

Create a table of values, and indicate the independent and dependent variable for the following:

A car's value decreases every year that it is driven. The Powell Motors Company briefly manufactured The Homer. The value of The Homer is given by the linear function:

$$V = -10250t + 82000$$

where t is the age of the car in years and V is the value of the car. Find the value of The Homer after 2 years, 5 years, and 8 years.

Based upon the description of the relationship:

Independent variable: time

Dependent variable: value of car

Time, t	2	5	8
Value, V	$V = -10250 \cdot 2 + 82000$ $= 61500$	$V = -10250 \cdot 5 + 82000$ $= 30750$	$V = -10250 \cdot 8 + 82000$ $= 0$

7.4 PRACTICE PROBLEMS

1. The price of gasoline is related to the cost of oil according to the equation

$$y = 2.8x + 60.9$$

where x is the cost of oil in dollars per barrel and y is the price of gasoline in cents per gallon. Evaluate the equation for various values of x and fill in the table below. Show your work.

Cost of Oil (dollars/ barrel)	Price of Gasoline (cents/gallon)
20	
40	
60	
80	
100	
120	

2. The amount of ice cream sales in a local shop is related to the temperature outside by the equation
$$y = 20.5x - 956$$

where y is the total sales (dollars) and x is the temperature in degrees Fahrenheit. Complete the table below:

Temperature (F°)	Total Ice Cream Sales ($)
58	
60	
54	
72	
75	
74	

3. Forced expiratory volume, FEV, measures the amount of air (in liters) that a person can forcibly exhale in one second. There is a linear relationship between the height of a person and their FEV:

$$y = 0.13x - 5.43$$

where x is height in inches and y is FEV in liters. Fill in the table below:

Height (in)	FEV (liters)
52	
48	
62	
74	
46	
70	

CHAPTER 8
INEQUALITIES, OPERATIONS, AND HYPOTHESIS TESTING

8.1 Order of Operations in the Test Statistic

Finding the numerical value of the test statistic in hypothesis testing is another challenging application of the order of operations. In these formulas there are several implied parentheses.

Consider the following examples.

EXAMPLE 1

Find the value of the test statistic when the proportion in the sample data, $\hat{p} = 0.60$, the proportion in the claim, $p = 0.50$ and the sample size, n, is 1000.

The formula for the test statistic is:

$$z = \frac{\hat{p} - p}{\sqrt{\dfrac{p \cdot q}{n}}} = \frac{(\hat{p} - p)}{\left(\dfrac{p \cdot q}{n}\right)^{\frac{1}{2}}}$$

The implied parentheses are explicitly given in the second part of this formula.

Step 1: Substitute into the formula. Be careful not to confuse p and $\hat{p}$. Note that since $p = 0.50$, then $q = 0.50$ because these must add up to 1.

$$z = \frac{(\hat{p} - p)}{\left(\dfrac{p \cdot q}{n}\right)^{\frac{1}{2}}} = \frac{(0.60 - 0.50)}{\left(\dfrac{0.50 \cdot 0.50}{1000}\right)^{\frac{1}{2}}}$$

Step 2: Evaluate the parentheses.

$$z = \frac{(0.60 - 0.50)}{\left(\dfrac{0.50 \cdot 0.50}{1000}\right)^{\frac{1}{2}}} = \frac{0.1}{(0.00025)^{\frac{1}{2}}}$$

Step 3: Perform the exponent operation.

$$z = \frac{0.1}{(0.00025)^{\frac{1}{2}}} = \frac{0.1}{0.0158}$$

135

Step 4: Finally, divide the numerator by the denominator.

$$z = 6.32 \text{ rounded to 2 decimal places}$$

EXAMPLE 2

Find the value of the test statistic when the mean in the claim μ, is 10 minutes, the mean of the sample, $\overline{x} = 12.5$ minutes, the sample size is 36, and the sample standard deviation s, is 2.6 minutes.

The formula for the test statistic is:

$$t = \frac{\overline{x} - \mu}{\frac{s}{\sqrt{n}}} = \frac{(\overline{x} - \mu)}{\left(\frac{s}{\sqrt{n}}\right)}$$

Step 1: Substitute.

$$t = \frac{(\overline{x} - \mu)}{\left(\frac{s}{\sqrt{n}}\right)} = \frac{(12.5 - 10)}{\left(\frac{2.6}{\sqrt{36}}\right)}$$

Step 2: Perform the innermost operation first then proceed to the next parentheses.

$$t = \frac{(12.5 - 10)}{\left(\frac{2.6}{\sqrt{36}}\right)} = \frac{(12.5 - 10)}{\left(\frac{2.6}{6}\right)} = \frac{2.5}{0.4333}$$

Step 3: Divide.

$$t = 5.769 \text{ rounded to three decimal places}$$

EXAMPLE 3

Find the value of the test statistic when the claimed standard deviation $\sigma = 15$, the sample size is 50, and the sample's standard deviation s, is 13.

The formula for the test statistic is:

$$\chi^2 = \frac{(n-1)s^2}{\sigma^2}$$

Step 1: Substitute.

$$\chi^2 = \frac{(n-1)s^2}{\sigma^2} = \frac{(50-1) \cdot 13^2}{15^2}$$

Step 2: Solve the parenthesis operation.

$$\chi^2 = \frac{(50-1) \cdot 13^2}{15^2} = \frac{49 \cdot 13^2}{15^2}$$

Step 3: Do the exponent operations.

$$\chi^2 = \frac{49 \cdot 13^2}{15^2} = \frac{49 \cdot 169}{225}$$

Step 4: Do multiplication and division.

$$\chi^2 = \frac{49 \cdot 169}{225} = \frac{8281}{225} = 36.8044 \text{ } \textit{rounded to four decimal places}$$

8.1 PRACTICE PROBLEMS

1. A survey was conducted to determine whether a majority of adults approve of the death penalty. Among the 1000 surveyed, 650 said that they favor the death penalty for certain crimes. Find the value of the test statistic with claimed proportion $p = 0.50$.

   a. Find the value of the sample proportion.

   $$\hat{p} = \frac{x}{n} =$$

   b. Identify each of the following variables.

   $\hat{p} =$
   $n =$
   $q =$

   c. Find the value of the test statistic. Show all of your work.

   $$z = \frac{\hat{p} - p}{\sqrt{\dfrac{p \cdot q}{n}}} =$$

2. One emergency care clinic claims that the mean waiting time is less than 10 minutes. In a sample of 40 visits to the clinic, the sample mean waiting time was 16.2 minutes with a standard deviation of 4.6 minutes.

   a. Identify the variables:

   $\overline{x} =$
   $n =$
   $\mu =$
   $s =$

   b. Find the value of the test statistic, t. Show all of your work.

   $$t = \frac{\overline{x} - \mu}{\dfrac{s}{\sqrt{n}}} =$$

3. Find the value of the test statistic when the claimed standard deviation $\sigma = 12$, the sample size is 30, and the sample's standard deviation s, is 18. Show all your work.

$$\chi^2 = \frac{(n-1)s^2}{\sigma^2} =$$

8.2 Interpreting Inequalities in Hypothesis Testing

Understanding inequalities is an important part of hypothesis testing. The inequalities are typically the result of interpreting a phrase used in a claim.

For example, consider the claim that a majority of voters approve of Referendum #1. Because majority means over half, the inequality would be $p > 0.5$. (The parameter p must be used here because majority also refers to a proportion over half.)

Review of Important Symbols

Symbol	Meaning	Equivalent Phrases
<	Less than	"fewer," "under," "below," or "smaller"
>	Greater than	"more," "over," "above," or "bigger"
≤	Less than or equal to	"At most" or "no more than"
≥	Greater than or equal to	"At least" or "no fewer than"
=	Equal to	"Is," "no different," "same as"

EXAMPLE 1

The Mars Co. claims that at least 15% of M&Ms are blue.

The parameter here is proportion or p, since it refers to a percentage. The phrase "at least 15%" means that amount, 15%, or any percentage greater than that amount. The symbol needed for this phrase is the "greater than or equal to" symbol, ≥.

The statement of this claim symbolically is:

$$p \geq 0.15$$

Note that the percent is customarily written as a decimal number.

EXAMPLE 2

RJ Reynolds Tobacco Co. claims that the mean nicotine amount in king-size cigarettes is no more than 1.1 mg.

The parameter here is the mean μ. The phrase "no more than 1.1 mg" includes 1.1 mg and any amount less than 1.1 mg. The symbol needed for this phrase is the "less than or equal to" symbol, ≤.

The statement of this claim symbolically is:

$$\mu \leq 1.1$$

EXAMPLE 3

Harvard College claims that the standard deviation of Harvard student's IQs is less than 15.

The parameter here is the standard deviation of a population σ. The phrase used is less than which would not include the number 15 only those numbers strictly less than 15.

The statement of this claim symbolically is:

$$\sigma < 15$$

8.2 PRACTICE PROBLEMS

Use the correct inequality to complete each of the symbolic claims below.

1. A bank claims that mean waiting time is less than 10 minutes.

$$\mu \ \boxed{} \ 10$$

2. A cigarette manufacturer claims that the mean amount of tar in its cigarettes is no more than 30 mg.

$$\mu \ \boxed{} \ 30$$

3. One candidate claimed that a majority of voters would vote for him.

$$p \ \boxed{} \ 0.5$$

4. A standardized test in math claims to have a standard deviation of 12 points.

$$\sigma \ \boxed{} \ 12$$

5. A local grocery claims that at least 53% of their customers use coupons.

$$p \ \boxed{} \ 0.53$$

8.3 Comparing Magnitude of Real Numbers for a Hypothesis Test

In order to make the decisions required in hypothesis testing, we need to be able to compare numbers properly.

In the *P*-value method, we are comparing the magnitude (size) of the *P*-value with the significance level, alpha (α). The critical value method requires that the critical value(s) and test statistic be placed on the number line relative to each other.

Determining the relative size of two numbers can be done by comparing the numerators of the fractions when they have same denominator. Similarly, two decimal numbers can be compared by aligning the decimal places and then, working from left to right, comparing the numbers in each place value position, until you reach a place where the numbers are different. The larger value in that place is the larger decimal number.

EXAMPLE 1

In running a hypothesis test, we find that the *P*-value is equal to 0.005 when the significance level, α, is 0.05.

The rule is that we reject the null hypothesis only when the *P*-value is less than or equal to the alpha level. So, we must compare the magnitude of these numbers.

Step 1: Make the numbers the same length.

P-value: 0.005 (which is the same as $\frac{5}{1000}$)

α: 0.050 (which is the same as $\frac{50}{1000}$)

Step 2: Compare the numbers.

$$0.05 > 0.005$$

So the *P*-value is less than the alpha significance level.

EXAMPLE 2

In a two-tailed test, we find that the test statistic equals 1.32 when the critical values are +1.96 and −1.96. We will reject the null hypothesis only if the test statistic is greater than 1.96 or less than −1.96.

To use the critical value method, we must be able to correctly place these numbers on the number line. This is the correct picture:

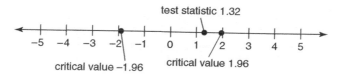

From the picture we can see that the test statistic is greater than −1.96 but less than 1.96. It is between these numbers.

8.3 PRACTICE PROBLEMS

In each problem compare the magnitude of the numbers using inequality symbols and the number line.

1. In a hypothesis test of whether a majority of voters would support Donald Trump for President, the P-value was found to be 0.0698 when $\alpha = 0.05$.

    a. $P\text{-}value$ ☐ α

    b. Graph the P-value and α on the number line provided below.

    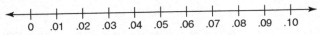

2. In a two-tailed hypothesis the claim was that 16% of M&Ms are red. The critical values for the test are ± 1.96 and the test statistic was calculated to be 3.26.

    a. Graph the test statistic and the critical values on the number line below:

    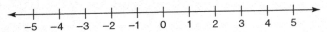

    b. Describe the sizes of these numbers using inequality symbols.

    $$-1.96 \ \boxed{} \ 3.26$$

    $$1.96 \ \boxed{} \ 3.26$$

3. In a hypothesis test of whether the mean waiting time to see an advisor was more than 10 minutes, the P-value was .0100 while the $\alpha = 0.05$.

    a. $P\text{-}value$ ☐ α

    b. Graph the P-value and α on the number line provided below.

    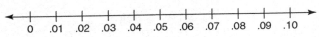

4. In a right-tailed test, the claim was made that the standard deviation of scores on the ACT math test for students entering two-year institutions is higher than that for students entering four-year institutions. In the test, the critical value was 24.996 while the test statistic was found to be 15.234.

    a. Graph the test statistic and the critical value on the number line below:

    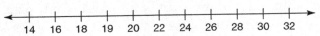

    b. Compare the size of these numbers using inequality symbols.

    $$24.996 \ \boxed{} \ 15.234$$

8.4 Graphing a Linear Equation Using Ordered Pairs

We often are asked to graph linear equations using ordered pairs. Remember that an ordered pair is in the format (x, y) so the first number is the x-coordinate and the second number is the y-coordinate.

For example, consider the linear equation $y = 2x + 3$. We know that the ordered pairs $(0, 3)$ and $(-1, 1)$ will satisfy this equation because when the x, y values from the coordinates are substituted into the equation, both pairs make the equation true.

To graph $y = 2x + 3$ we begin by graphing the location of both points. The plot below shows the location for the points $(0, 3)$ and $(-1, 1)$:

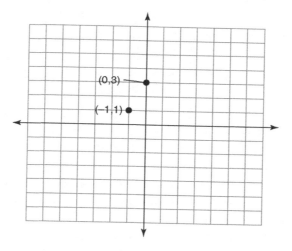

To complete the graph of the line, we need only connect these points with a straight line:

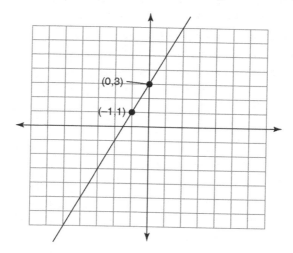

8.4 PRACTICE PROBLEMS

Use the graph paper below to graph each linear equation. Label each line and use a different color pencil for each one. Create a table of last least two points which satisfy each equation and use them to graph the equation.

1. $y = -3x + 1$

x	y
−1	
0	
1	

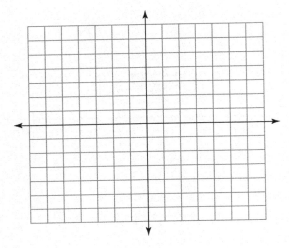

2. $y = -1 + \dfrac{1}{2}x$

x	y
−1	
0	
1	

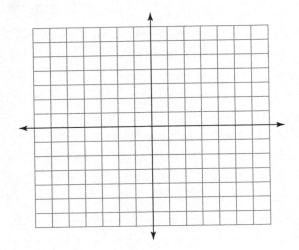

3. $y = \dfrac{1}{3}x + 2$

x	y
−1	
0	
1	

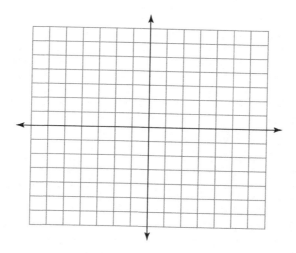

UNIT V

PRINCIPLES OF INFERENCE FOR PAIRED AND QUALITATIVE DATA

9.1 Identifying Slope and *y*-intercept of a Linear Equation

A linear equation is used to describe the graph of a straight line in the Cartesian plane. By definition, slope describes the rate of change of the *y* values relative to the *x* values. The *y*-intercept represents the *y*-value of the point where the line crosses the vertical axis. This number can be either positive or negative.

Slope-intercept Form

To determine the slope and *y*-intercept of a linear equation, the form that we will find most useful is given by the literal equation below:

$$y = mx + b$$

In this form, *x* and *y* are the coordinates of any ordered pair that are on the line, *m* is the slope of the line and *b* is the *y*-intercept.

When the line is going up (or increasing) from left to right, the line has a positive slope and *m* is a positive number. When the line is going down (or decreasing) from left to right, the line has a negative slope and *m* is a negative number. The *x*-coordinate of the *y*-intercept is always 0. Therefore the ordered pair for the *y*-intercept can be written (0, *b*).

In the slope-intercept equation below, the value of *b* is 3. This means the line crosses the vertical axis at the coordinates (0, 3). Examine the graph below for the illustration.

$$y = 2x + 3$$

y intercept
(0,3)

Let's look at some examples of how we would identify the slope and *y*-intercept for linear equations.

EXAMPLE 1

Identify the slope and *y*-intercept for the equation, $y = -2x + 5$.

This equation is already in slope-intercept form so we simply inspect the equation. The slope is the coefficient of the *x* variable, so $m = -2$. The *y*-intercept is the constant term, $b = 5$. This can also be written as the ordered pair $(0, 5)$.

Equations not in Slope-intercept form

If a linear equation is not written in slope-intercept form, you must first solve for *y* (by isolating the *y* variable) in order to identify the slope and *y*-intercept values. Once you have solved a linear equation for *y* then the slope is the coefficient multiplying the *x* variable and the *y*-intercept is the constant term.

EXAMPLE 2

Identify the slope and *y*-intercept for the equation, $3x - 3y = 6$.

Step 1: We need to put this equation into slope-intercept form by solving for *y*.

Subtract $3x$ from both sides.

$$3x - 3x - 3y = 6 - 3x$$
$$-3y = -3x + 6$$

Divide each term on both sides by -3.

$$\frac{-3y}{-3} = \frac{-3x}{-3} + \frac{6}{-3}$$
$$y = x - 2$$

Step 2: Identify the slope and the *y*-intercept. By inspection of the equation:

$$m = 1$$
$$b = -2 \text{ or } (0, -2)$$

EXAMPLE 3

Identify the slope and *y*-intercept for the equation, $2x - y = 7$.

Step 1: Convert to slope-intercept form.

Subtract $2x$ from both sides.

$$2x - 2x - y = 7 - 2x$$
$$-y = 7 - 2x$$

Divide each term on both sides by -1.

$$\frac{-y}{-1} = \frac{7}{-1} - \frac{2x}{-1}$$
$$y = -7 + 2x$$

Step 2: Identify the slope, m, and the y-intercept. By inspection of the equation:

$$m = 2$$
$$b = -7 \text{ or } (0, -7)$$

9.1 PRACTICE PROBLEMS

Identify the slope and the intercept for each of the problems below. If necessary, solve the equation for y so that the equation is in slope-intercept form.

1. $y = 5x - 1$

2. $y = \dfrac{3}{2}x + 5$

3. $3x - 2 = y$

4. $2x + 4y = 13$

5. $14x - y + 2 = 0$

6. $\dfrac{1}{2}y + 3x = 4$

7. $2x + 6 = y - 3$

8. $4x - 2 = y$

9.2 Interpreting Slope as a Rate of Change

If you look at a graph of a line the idea of slope is the amount of rise (vertical direction change) over the amount of run (horizontal direction change) between two points. The vertical direction change is the difference in the y-coordinates and the horizontal change is the difference in the x-coordinates.

We can express this relationship as an equation for the slope m between two points (x_1, y_1) and (x_2, y_2):

$$m = \frac{rise}{run} = \frac{y_2 - y_1}{x_2 - x_1}$$

Consider the examples below.

EXAMPLE 1
What is the slope of a line that contains the points (2, 14) and (5, 20)?

Step 1: The slope formula is:

$$m = \frac{rise}{run} = \frac{y_2 - y_1}{x_2 - x_1}$$

We need to identify the x- and y-values for each point.

- Let the first point (2, 14) be (x_1, y_1) so that

$$x_1 = 2 \text{ and } y_1 = 14.$$

- Then the second point (5, 20) would be (x_2, y_2) so that

$$x_2 = 5 \text{ and } y_2 = 20.$$

Step 2: Substitute into the equation:

$$m = \frac{y_2 - y_1}{x_2 - x_1} = \frac{20 - 14}{5 - 2} = \frac{6}{3} = 2$$

In this particular case the vertical direction change (or rise) is 2 and the horizontal direction change (or run) is an implied 1. This indicates that the line is "rising" twice as fast as it is "running."

A slope of 2 looks like this graphically:

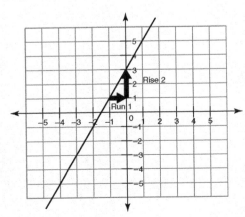

EXAMPLE 2

Calculate the slope of the line containing points (3, 8) and (5, 9).

Step 1: The slope formula is:

$$m = \frac{rise}{run} = \frac{y_2 - y_1}{x_2 - x_1}$$

- Let the first point (3, 8) be (x_1, y_1) so that

$$x_1 = 3 \text{ and } y_1 = 8.$$

- Then the second point (5, 9) would be (x_2, y_2) so that

$$x_2 = 5 \text{ and } y_2 = 9.$$

Step 2: Substitute into the equation:

$$m = \frac{y_2 - y_1}{x_2 - x_1} = \frac{9 - 8}{5 - 3} = \frac{1}{2}$$

Note that in this case the line is changing vertically 1 unit for every horizontal change of 2 units. This slope would look like the following graph:

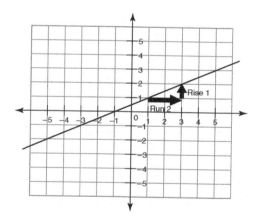

EXAMPLE 3

Calculate the slope of the line containing points (8, 14) and (11, 12).

Step 1: The slope formula is:

$$m = \frac{rise}{run} = \frac{y_2 - y_1}{x_2 - x_1}$$

- Let the first point (8, 14) be (x_1, y_1) so that

$$x_1 = 8 \text{ and } y_1 = 14.$$

- Then the second point (11, 12) would be (x_2, y_2) so that

$$x_2 = 11 \text{ and } y_2 = 12.$$

Step 2: Substitute into the equation:

$$m = \frac{y_2 - y_1}{x_2 - x_1} = \frac{12 - 14}{11 - 8} = \frac{-2}{3}$$

Note that in this case the line has a vertical change of −2 (going down) for every horizontal change of 3. This slope would look like the following graph:

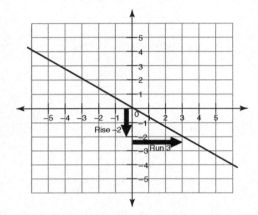

The slope of a line can be interpreted as a rate of change, or the amount the dependent y-values change relative to the independent x-values.
Consider the example below.

EXAMPLE 4

I pay a flat fee of \$4.99 per month and \$0.25 per minute for international calls with my telephone service provider. This can be expressed as the linear equation, $y = 0.25x + 4.99$.

Describe what this means in terms of the rate of change of the dependent variable per unit change in the independent variable.

The dependent variable y is my total monthly cost and the independent variable x is the number of international minutes used that month.

The slope in this equation is 0.25 which tells us that my total monthly cost increases by \$0.25 for every international minute used.

9.2 PRACTICE PROBLEMS

For each problem (1–4), determine the slope as defined by the equation:

$$m = \frac{y_2 - y_1}{x_2 - x_1}$$

1. Point 1: (0, 3)

   Point 2: (5, 2)

$$y_2 - y_1 =$$
$$x_2 - x_1 =$$
$$m =$$

2. Point 1: (−1, 2)

   Point 2: (−6, 4)

3. Point 1: (1, 1)

   Point 2: (4, 4)

4. Point 1: (5, 0)

   Point 2: (5, 1)

5. The cost C, of a medium cup of frozen yogurt is \$3.25 plus \$0.30 for each additional topping x, added. This can be represented by the linear equation

$$C = 3.25 + 0.30x$$

   a. What are the independent and dependent variables?

b. What is the minimum cost (y-intercept) for a medium cup of frozen yogurt?

c. Describe what the slope means in terms of the rate of change of the dependent variable per unit change in the independent variable.

6. The number of students enrolled at a community college each year after 2009 can be modeled by the linear equation

$$E = 12500 - 100t$$

a. What are the independent and dependent variables?

b. What is the maximum enrollment (y-intercept) of students at the college?

c. Describe what the slope means in terms of the rate of change of the dependent variable per unit change in the independent variable.

9.3 Determining the Equation of a Line

Given specific information about the variables, we can write the equation of the line in slope-intercept form, $y = mx + b$.

We must first identify the independent variable x, and the dependent variable y. Next, we also need to examine the description to determine the slope (how the y-values are changing relative to the x-values). Finally, we need to determine the constant y-intercept value.

Consider the examples below.

EXAMPLE 1

What is the equation for a line which crosses the vertical axis at (0, 5) and goes up 2 as it goes over 3?

This description is telling us that the slope is $\frac{2}{3}$ because slope is defined as the change in the vertical direction over the change in the horizontal direction (or rise over run).

In addition, the y-intercept is defined as the y-value where the line crosses the vertical axis. So the y-intercept here is 5.

The equation is:

$$y = \frac{2}{3}x + 5$$

EXAMPLE 2

The starting salary for a particular industry is $28000 plus $2500 for each additional year of education beyond high school. Write the equation of the line for starting salaries.

The independent variable is years of education beyond high school t, and the dependent variable is starting salary, S.

The minimum starting salary with 0 years of education beyond high school is $28000. Since the value of the independent variable is 0, this is our y-intercept.

The slope describes how the salary changes based on years of education beyond high school. So the slope here is 2500, since each year corresponds to a $2500 increase.

The equation would be:

$$S = 2500t + 28000$$

9.3 PRACTICE PROBLEMS

1. What is the equation for a line which crosses the vertical axis at $(0, 0)$ and goes up 4 as it goes over 5?

2. What is the equation for a line which crosses the vertical axis at $(0, -2)$ and goes up -3 as it goes over 2?

3. What is the equation for a line which crosses the vertical axis at $(0, 4)$ and goes up 1 as it goes over 3?

4. A certain company makes swizzle sticks. Each package of sticks cost $1.00 in materials and overhead costs are a flat amount of $2000 per month.

   a. What equation represents the company's costs per month?

   b. What value is the y-intercept?

   c. What value is the slope?

5. An apartment complex has a large party room which can be rented for a flat fee of $100 plus $10 per person at the party to have small snack foods catered. Let y represent the total costs to have a party.

   a. What equation expresses total cost when x represents the number of people at the party?

   b. If the party is cancelled ($x = 0$), the room rental fee is still owed. What does this value represent?

   c. What value represents the slope, the change in the cost based on the number of people attending the party?

   d. How much do is owed if 50 people go to the party? Show your work.

9.4 Graphing a Linear Equation Using the *y*-intercept and Slope

While we sometimes graph straight lines by creating a table of values and plotting individual points, we often find it easier to simply work from the linear equation in slope-intercept form, $y = mx + b$.

Consider the following examples.

EXAMPLE 1

Graph the equation $y = 2x + 3$.

Step 1: We begin the graph at the *y*-intercept, the place where the line crosses the *y*-axis. The *x*-value here would have to be 0. The *y*-value of this point comes from the value of *b* in the equation which is 3. In this case, that point is (0, 3).

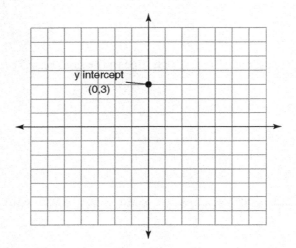

Step 2: From this point we need to use the slope to move to another point on the line.

$$m = \frac{rise}{run} = \frac{2}{1}$$

Therefore, we need to rise positive 2 units from our starting point while running positive 1 units over.

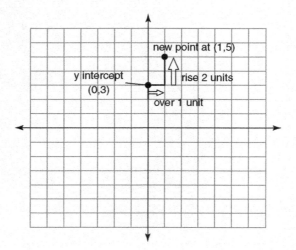

Connecting these points with a straight line will complete our graph of the line $y = 2x + 3$.

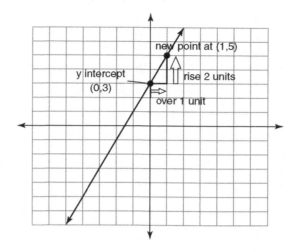

EXAMPLE 2

Graph the equation $2x + y = 4$.

Step 1: First, convert to slope-intercept form by solving for y.

$$2x + y - 2x = 4 - 2x$$

$$y = -2x + 4$$

Step 2: Identify the slope and the y-intercept. By inspection of the equation:

$$m = -2 = \frac{-2}{1} = \frac{rise}{run}$$
$$b = 4 \text{ or } (0, 4)$$

Step 3: Plot the y-intercept.

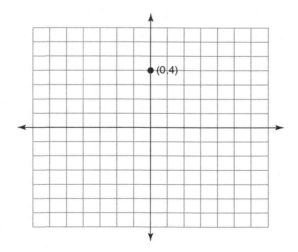

Step 4: Use the slope to move to another point on this line. Since the slope is −2, we will go down (the negative direction) by two and then over positive one.

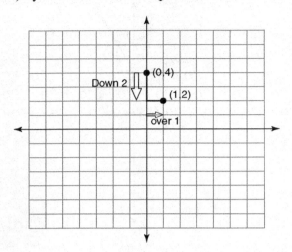

Step 5: Connect these points with a line and extend in both directions.

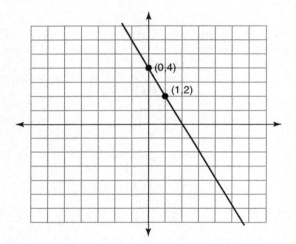

9.4 PRACTICE PROBLEMS

Use the graph paper below to graph each linear equation. Label each line and use a different color pencil for each one.

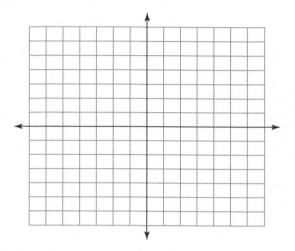

1. $y = -3x + 1$

2. $y = -1 + \dfrac{1}{2}x$

3. $y = \dfrac{1}{3}x + 2$